建筑立场系列丛书 No.60

C3

地方性与全球多样性
Regionalism and Global Diversity

汉英对照
(韩语版第376期)

韩国C3出版公社 编

张琳娜 时跃 蒋丽 陈帅甫 译

大连理工大学出版社

建筑立场系列丛书 No.60

4
- 004 2015青年建筑师国际项目——MoMA
- 005 纽约，MoMA PS1：COSMO _ Andrés Jaque/Office for Political Innovation
- 010 首尔，MMCA：屋顶情感 _ Society of Architecture
- 016 罗马，MAXXI：大地 _ CORTE
- 018 伊斯坦布尔，伊斯坦布尔现代馆：所有坚固的东西 _ PATTU

20 地方性与全球多样性
- 020 地方与全球 _ Aldo Vanini
- 026 帕尼社区中心 _ SchilderScholte Architects
- 038 丘基班比亚学校 _ AMA-Afonso Maccaglia Architecture + Bosch Arquitectos
- 046 健康和社会进步中心 _ Kéré Architecture
- 056 Thread艺术家住所与文化中心 _ Toshiko Mori Architect
- 070 牛背山志愿者之家 _ dEEP Architects
- 080 漂浮在天空中的孤儿学校 _ Kikuma Watanabe
- 090 Trika别墅 _ Chiangmai Life Construction
- 102 紧凑的喀斯特房屋 _ Dekleva Gregorič Arhitekti
- 114 P住宅 _ Budi Pradono Architects

128
住宅
居住在记忆里
游走在复兴与再利用之间
- 128 居住在记忆里：游走在复兴与再利用之间 _ Angelos Psilopoulos
- 134 奥洛特某联排住宅 _ RCR Arquitectes
- 148 黛波拉某独栋住宅 _ Irisarri-piñera S.L.P.
- 162 Scaiano旧石屋改造项目 _ Wespi de Meuron Romeo Architects
- 176 花园住宅 _ Hertl. Architekten ZT GmbH

- 184 建筑师索引

4

004 2015 Young Architects Program International – MoMA

005 New York, MoMA PS1: COSMO _ Andrés Jaque/Office for Political Innovation

010 Seoul, MMCA: Roof Sentiment _ Society of Architecture

016 Rome, MAXXI: Great Land _ CORTE

018 Istanbul, Istanbul Modern: All that is Solid _ PATTU

20

Regionalism and Global Diversity

020 *Local vs. Global _ Aldo Vanini*

026 Pani Community Center _ SchilderScholte Architects

038 Chuquibambilla School _ AMA-Afonso Maccaglia Architecture + Bosch Arquitectos

046 Center for Health and Social Advancement _ Kéré Architecture

056 Thread Artist Residency and Cultural Center _ Toshiko Mori Architect

070 Cattle Back Mountain Volunteer House _ dEEP Architects

080 Floating in the Sky School for Orphans _ Kikuma Watanabe

090 Trika Villa _ Chiangmai Life Construction

102 Compact Karst House _ Dekleva Gregorič Arhitekti

114 P House _ Budi Pradono Architects

Dwell How

128

Dwelling in Memory

between rehabilitation and reuse

128 *Dwelling in Memory: between rehabilitation and reuse _ Angelos Psilopoulos*

134 Row House in Olot _ RCR Arquitectes

148 Single Family House in Tebra _ Irisarri-piñera S.L.P.

162 Old Stone House Conversion in Scaiano _ Wespi de Meuron Romeo Architects

176 Garden House _ Hertl. Architekten ZT GmbH

184 Index

2015青年建筑师国际项目——MoMA

第十六届青年建筑师项目在MoMA（纽约现代艺术博物馆）和MoMA PS1举行。项目致力于为刚刚崭露头角的建筑师提供机会设计和展示具有创新意义的项目，使建筑师每年为位于MoMA PS1的室外临时性设施进行创意设计并争夺冠军，该设施可提供遮阳、休息和饮用水。建筑师们必须遵循有关环境问题的指导方针，如可持续发展和再循环利用。该项目的目的是充分利用原有的空间，就地取材，为游客在城市环境中创建一个遮蔽夏季炎炎烈日的户外休闲场所。项目预算有限，建筑师参与到了项目设计、开发和建设的每一个方面。MoMA PS1宽敞的三角形入口庭院和户外雕塑区是博物馆以实验音乐、现场乐队和DJ为特色的流行音乐系列音乐会Warm Up的主要组成部分。场地整个夏天都对游客开放。

MoMA和MoMA PS1与意大利罗马的21世纪艺术博物馆（MAXXI）、智利圣地亚哥的CONSTRUCTO博物馆、土耳其伊斯坦布尔的现代博物馆（双年展）还有韩国首尔的MMCA博物馆联合创办了YAP国际展。

为了在MoMA PS1评选出竞选YAP的建筑设计公司，建筑学校的校长和建筑出版物编辑推荐了由在校生、刚刚毕业的建筑学校毕业生和已经注册的建筑师成立的30个公司，共同研究与尝试新的风格和技术。这些团体需要提交作品文件以供审阅。评审团最终评选出五个参赛团体，这五个参赛团体将受邀设计指定场所的最初提案。

2015 Young Architects Program International

Now in its 16th edition, the Young Architects Program(YAP) at MoMA and MoMA PS1 has been committed to offering emerging architectural talent the opportunity to design and present innovative projects, challenging each year's winners to develop creative designs for a temporary, outdoor installation at MoMA PS1 that provides shade, seating, and water. The architects must also work within guidelines that address environmental issues, including sustainability and recycling. The objective of the project is to provide visitors with an outdoor recreational area for the summer – a much-needed refuge in an urban environment – making the best use of the pre-existing space and available materials. The architects follow a program with a tight budget, and are involved in every aspect of the design, development, and construction of the project. The site, MoMA PS1's large triangular entrance courtyard and outdoor sculpture area, is an integral part of the museum's popular music concert series, Warm Up, which features experimental music, live bands, and DJs. The site is open to visitors throughout the summer.

MoMA and MoMA PS1 have partnered with the National Museum of XXI Century Arts (MAXXI) in Rome, Italy, with CONSTRUCTO in Santiago, Chile, and with Istanbul Modern in Istanbul, Turkey (on a biennial cycle) and with MMCA in Seoul, Korea to create international editions of the YAP.

纽约，MoMA PS1，2014年——Hy-Fi, The Living
New York, MoMA PS1, 2014 _ Hy-Fi, The Living

首尔，MMCA，2014年——ShinSeon Play, Moon Ji Bang
Seoul, MMCA, 2014 _ ShinSeon Play, Moon Ji Bang

纽约——MoMA PS1
New York _ MoMA PS1

首尔——MMCA
Seoul _ MMCA

罗马——MAXXI
Rome _ MAXXI

伊斯坦布尔——伊斯坦布尔现代馆
Istanbul _ Istanbul Modern

To choose an architectural firm for the YAP at MoMA PS1, deans of architecture schools and the editors of architecture publications nominate some 30 firms comprised of students, recent architectural school graduates, and established architects experimenting with new styles or techniques. The group is asked to submit portfolios of their work for review. The panel selects five finalists who are invited to make preliminary proposals for the designated site.

纽约，MoMA PS1: COSMO_Andrés Jaque/Office for Political Innovation

给我一根管道，我将移动整个地球

COSMO作为获胜项目自2015年6月下旬在长岛市MOMA PS1进行展出。纽约市地下每天有20亿加仑以上的水在循环。COSMO是一个由特制的灌溉组件组成的可移动手工制品，它使我们赖以生存的隐形城市环境中的管道赏心悦目地呈现在人们眼前。一整套以先进的环境设计为基础的生态系统对3000加仑水进行过滤、净化，去除悬浮颗粒和硝酸盐，平衡PH值，提高溶解氧水平。

据联合国估计，到2025年，全球三分之二的人口将生活在缺少充足水源的国家。COSMO被设计成既是离线的也是在线的原型。其目的在于激发人们的节水意识，使全世界范围内的水能通过简便的再生产方式来获取，从而使人们得到饮用水，同时引发人们关于水的再利用的对话。但最重要的是，COSMO是一个为聚会而制的手工制品，有聚会的地方就有它。它是一个旨在将人们聚集在一起的装置，如同花园般让人心旷神怡，气候怡人，同时又像一个迪斯科舞会般熠熠生辉。由于运用了复杂的生物化学设计，水一旦被净化，装置外部的拉伸塑料网就会自动闪光。环境每一次得到保护，COSMO都会点亮聚会。

New York, MoMA PS1: COSMO

Give me a pipe and I will move/celebrate the earth

The winning project, COSMO has been exhibiting at MoMA PS1 in Long Island City from late June. More than 2 billion gallons of water circulate everyday beneath New York City. COSMO is a movable artifact, made out of customized irrigation components, to make visible and enjoyable the so-far hidden urbanism of pipes we live by. An assemblage of ecosystems, based on advanced environmental design, engineered to

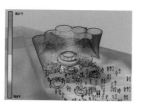

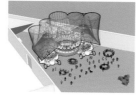
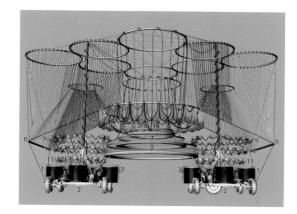

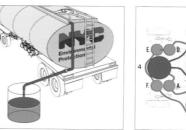

Newton creek water treatment plant

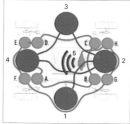

1. open aerobic reactor
2.3.4. closed anaerobic reactors
5. water pump
A.B.C.D. clarifier artificial wetland
E.F.G.H. floating wetland

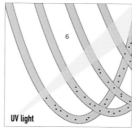

6. transparent hose
UV light

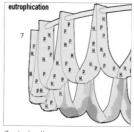

7. algal cells
H. hydrogen
P. phosphorus

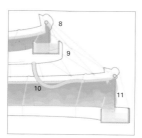

8. perforated pipe
9. perimeter gutter
10. connecting hose
11. algal turf screens

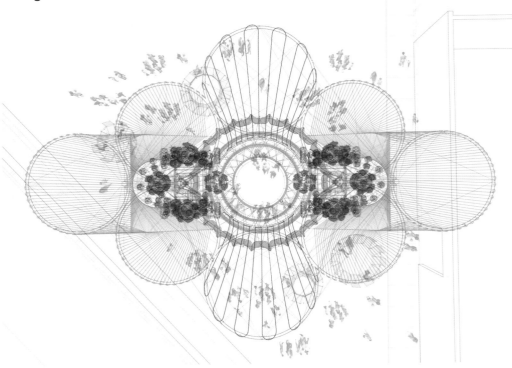

a. non organic particles b. peat
沼泽 bog

a. water bodies
湿地 wetland meadows

悬浮的根部 suspended roots

filter and purify 3,000 gallons of water, eliminating suspended particles and nitrates, balancing the PH, and increasing the level of dissolved oxygen.

The United Nations estimates that by 2025, two thirds of the global population will live in countries that lack sufficient water. COSMO is designed as both an off-line and an on-line prototype. Its purpose to trigger awareness, to be easily reproduced all around the world, giving people access to drinking water, and to a dialogue about it. But above all, COSMO is a party artifact that moves to go there wherever the party happens. It is a device meant to gather people together, as pleasant and climatically comfortable as a garden and at the same time as visually rich as a mirrored disco ball. As a result of a complex biochemical design, its stretched out plastic mesh glows automatically whenever its water has been purified. With COSMO, the party is literally lit up every time when the environment is being protected.

首尔，MMCA：屋顶情感_Society of Architecture

2015青年建筑师国际项目 2015 YAP International

过去，屋顶应该是巨大而又沉重的，用来加固木柱。屋顶曾经非常引人注目，因为它与天空紧密相连。屋顶将天气遮挡在外。屋顶之下，我们感受得到高大而深邃的空间与气氛、凉爽的微风、暗淡的光线和人们的身影。

随着技术的发展，屋顶变得更加轻薄。由于屋顶变得又薄又平，于是失去了"下层的空间"以及所有与之相关的情感。

平屋顶堆叠起来，建筑的层数增加，使房屋失去了屋顶情感。屋顶上方是另一层楼板而非天空。在无形的平屋顶之下，我们缺乏那些凸显屋顶最初品质的技巧。

我们在MaDang（院子）里设计了这个具有褶皱的屋顶。它是一个能观赏风景的窗口，也是一个能遮风的百叶窗，这个波浪形的装置能够唤醒人们的屋顶情感。这个屋顶不是覆盖在下面的空间之上，而是触动了人们的感觉和情感。

屋顶的褶皱松散地划分开下面的空间。人们在屋顶下占据各自的空间。屋顶将狭小的空间连接在一起，让人们能够走进去。傍晚，屋顶升起时，人们可以聚在一起友好地交谈，有时也举行小型的夏日庆祝活动。与将天气隔在外面不同，这个屋顶能使人们感受到天气，并与自然交谈。人们透过屋顶的层次感受彼此间的深意。在摇摆的屋顶下，人们在夏日清风里野餐，观看表演。因为屋顶的摇摆，每日的例行程序开始有出乎意料的改变。人们对随风摇摆的屋顶做出回应。

Seoul, MMCA: Roof Sentiment

In the past, roof should be large and heavy to reinforce the wooden pillars. Roof used to be outstandingly visible, due to the closest connection to the heavens. Roof kept the weather out. And under the roof we experienced high and deep space, atmosphere, cool breeze, dim light, and shadow figure. With technology development, roof becomes to be more thinner and lighter. As roof becomes to be thin and flat, roof has lost "under-space" and all of these sentiments.
Flat roofs were stacked up and the stories of building grew, so house lost the roof sentiments, and the roof-over-our-head

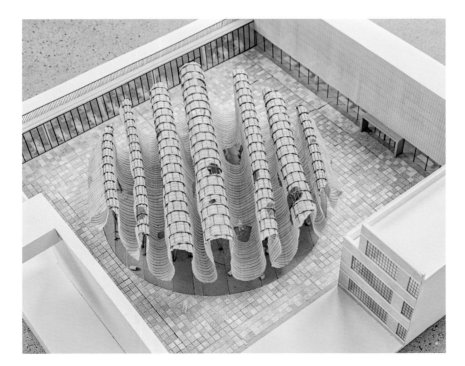

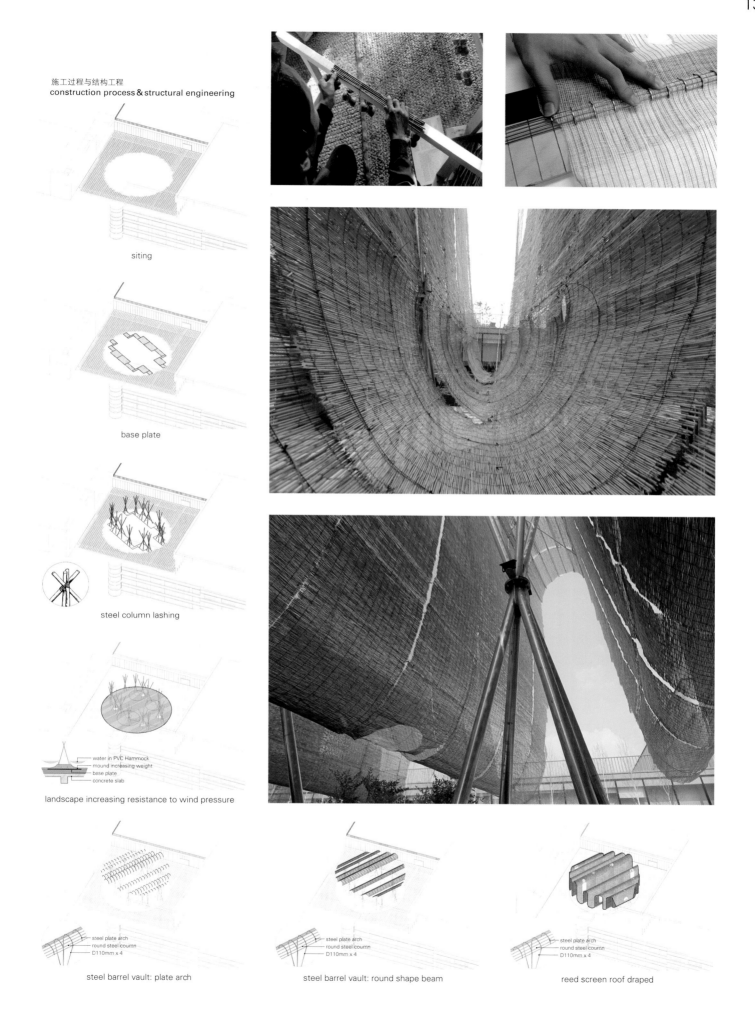

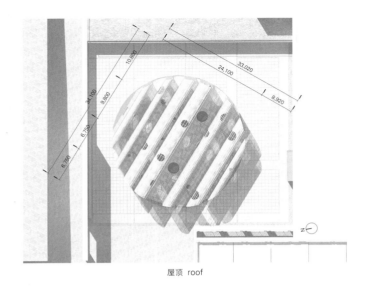

屋顶 roof

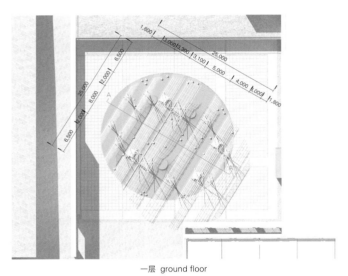

一层 ground floor

stand at the other floor instead of sky. Under the flat and invisible roof, we lack the skills to articulate the roof's primal qualities.

We propose the wrinkle roof in this MaDang(yard). This roof is a window to borrow scenery, a blind to protect against wind, and a wavy installation to awaken people the roof sentiment. This roof is not covering the under space, but uncovering people's feeling and sense.

The wrinkles of the roof divide the underneath loosely. Under the roof people occupy their own place. The roof connect the small space together, and make people get into it. When the roof is up, people gather around together and have a friendly talk in the evening, sometimes a small summer festival happens. Instead of keeping the weather out, roof makes people feel the weather and chat with the nature. People see through the roof layers and start feeling the depth between you and I. Underneath the swaying roof, picnics in summer wind and performances happen. As the roof sways, everyday routine starts to change unexpectedly. People react to the swaying roof with breeze. Society of Architecture

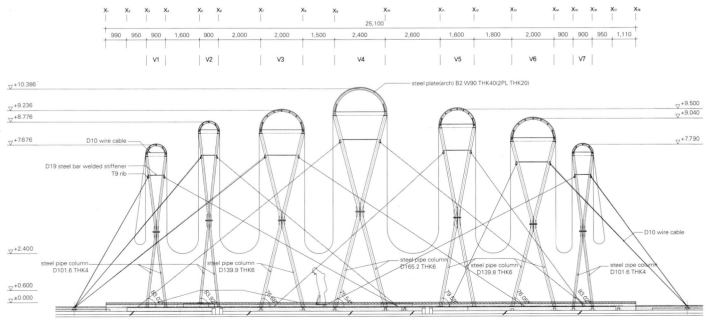

A-A' 剖面图 section A-A'

罗马，MAXXI：大地_CORTE

项目"大地"由一大片草地和山丘交替的土地组成；它是一个长30m、宽9m、高4m的平行六面体，表面覆盖的反射层更增添了无限的风景。草坪种植的植物都是乡村自然生长的典型植被，还有五颜六色的野餐桌布。部分桌布是固定的，其中隐藏着喷水口，草坪上还有一些设施可供游客使用。项目的一侧设有举办夏日YAP活动节的舞台。

CORTE事务所的项目最终获选是因为建筑师将抽象的项目清晰化，使其与呈现给城市的文化公园的实用性相结合。

"大地"同时利用野餐区和绿地与当前MAXXI正在举行的"Food dal cucchiaio al mondo"食物社会空间展的主题相连接，巩固了空间价值，这种空间价值包含在概念建筑的简单性、小型景观的柔和性以及艺术的直接性之内。

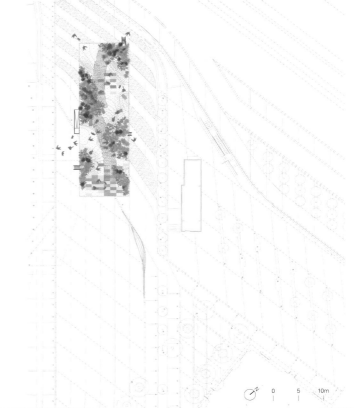

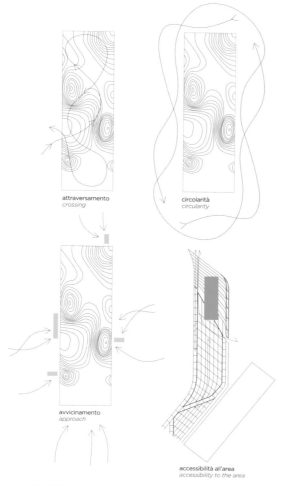

WHERE

The clod of soil stands at the center of the esplanade, isolated from the perimeter of the area and indifferent to altitude changes. A thickness distinguishes it in relation to the context, to remark the existence of an external world and an internal one inside the project. The first is dedicated to the discovery of the scene, and the second is based on the experience and on the exploration of the finally conquered territory.

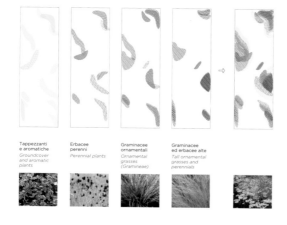

HOW

OROGRAPHY

An alternation of hills and flat parts draws the landscape and suggests different uses. Entertainment and dynamism in the stage or water feature area relaxation between the more intimate valleys.

VEGETATION

The taste is rustic, spontaneous, exuberant. The vegetation becomes progressively more sparse in flat areas, eroded by the man action. The margin between the grass and the wild area is uncertain and variable.

Rome, MAXXI: Great Land

Great Land consists of a large clod of earth with alternating meadows and hills; it is a parallelepiped that's 30 meters long, 9 meters wide, and 4 meters high, clad with a reflecting surface that multiplies the landscape within it infinitely. The lawn is strewn with plants that are typical of the vegetation growing spontaneously in the countryside, as well as colorful picnic tablecloths. Some of them are fixed and hide nozzles from which water spurts out, and there is furniture made available to visitors as well. On one side of the installation is a stage hosting the events of the summer YAP Festival.

CORTE's project was chosen for its capacity to match the clarity of a project with a high rate of abstraction with the utilitarian quality of a cultural park offered to the city.

With its picnic areas and greenery, Great Land is also linked to the theme of the social space of food dealt with by the exhibition "Food dal cucchiaio al mondo" currently underway at MAXXI, and it consolidates the value of a space that contains within the simplicity of the conceptual architecture, the gentleness of a small landscape, and the immediacy of art.

2015青年建筑师国际项目 2015 YAP International

伊斯坦布尔,伊斯坦布尔现代馆:所有坚固的东西_PATTU

PATTU事务所声称,"所有坚固的东西"的设计灵感源于该地区的工业历史,而建造该建筑的目的不仅仅是为了纪念过去,也是在陈述未来所承载的即将发生的改变,如此一来我们就对它更加挑剔:"建筑就好像是人类回忆的集合体,因为它们往往比我们人类的寿命要长久许多。但是它们到底有多坚固呢?它们是由石料、水泥和大理石材料制成的,然而这一事实并不能真正使它们永恒存在。有可能它们在一天之内就会消失殆尽。所以,我们借用了名言的一部分"所有坚固的东西都烟消云散了"来表达我们的设计理念。我们的设计方法是通过其所有的构件和先前的建筑来仔细分析伊斯坦布尔现代馆周围的空间,并且将它们重新集合,以展示建筑转瞬即逝的一面。

设计借用过去曾经位于该区域的建筑的几何形状,并且以一种无序的方式将它们集合到一起。但当建筑的透明形状变得不透明时,这种无序混乱随着一天中时间的推移,开始逐渐变得合乎情理。过去的几何形状变得明显,同时也更加不明显。

Istanbul, Istanbul Modern: All that is Solid

According to PATTU, who were inspired by the industrial history of the area, "All that is Solid" is not just a reminder of the past but also a statement about the imminent change the

西立面 west elevation

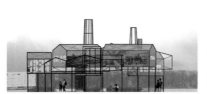
南立面 south elevation

东立面 east elevation

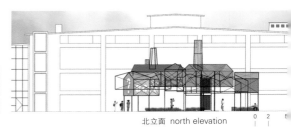
北立面 north elevation

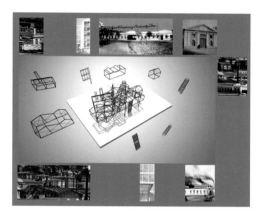

future holds, so that we can be more critical about it: "Buildings are like anchors to memories because they tend to last longer than our human lifespan. But how solid are buildings? The fact that they are made of stone, cement and marble does not really make them last eternally. They can easily be gone in a day. So we borrowed a part of the famous quote "All that is solid melts into air." Our approach was to dissect the space around Istanbul Modern with all its elements and previous constructions, and reassemble them, showing the ephemeral side of architecture.

The design borrows geometries from past buildings that once stood in the area and crunches them together in a chaotic way. But this chaos starts making sense over the course of a day as the transparent shapes become opaque. Past geometries become visible, and more invisible. PATTU

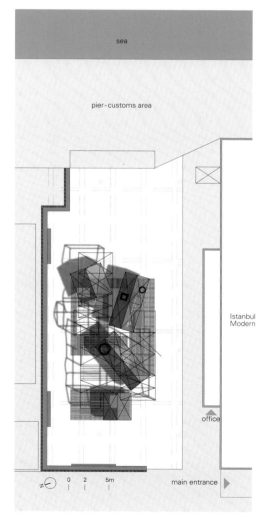

地方性与全球多样性

Regionalism and Global

经济和文化全球化势不可挡的影响力,作为当代每项活动的统一模式,产生了一种具有干扰性的理念——"全球化的"便是现代和时尚的,而"地方性的"则是过时的。

市场对决策、资源使用、最终的文化选择以及现在的商品化的影响力与日俱增,使得各个阶层的决策人保持着循规蹈矩的态度,最终形成一般化且平淡无奇的结果。而地方社区则将这种形式主义的全球性运动中最具代表性的对象看作是外星生物,它们最大的用处是作为地标或供参观者和游客观光的景区。

另一方面,如果在对全球性同化现象做出反应时忽略了根源以及当地文化和传统的影响,就很容易产生或保守主义(最好的情况)或取悦于怀旧和地方情形(最坏的情况)的观点。在回答建筑是否在每个时间里都能够适应特殊的气候和地理条件,甚至可以开发出与当地环境相关的特殊建造技术这样的问题时,以上这两种情况都使得我们思考问题中所涵盖事物的丰富性和复杂性。公众认为的遵循国际套路的现代性与空间和建筑的地方文化之间的对比,其实是一个错误的并且不恰当的问题。地方性建筑传统只表面看起来是静止的,无论如何,它都是历史不断演进的结果,而正是因为这一原因,它能够并且必定产生其自身对现代性的表达。而关注由地方性经验,或至少由外部经验和地方特色之间的碰撞产生的建筑风格,对克服全世界范围内的壮观纪念碑类型缺失问题来说也是至关重要的。

建筑学的新型地方性解决方案并不是民间传统的怀旧考古学,而是与异常丰富的当地文化密切相关的当代空间和人类学概念。

正如尊重生物多样性以确保生存和自然环境稳定性一样,尊重文化多样性保证了进化型建筑思想的存在。

帕尼社区中心_Pani Community Center/SchilderScholte Architects
丘基班比亚学校_Chuquibambilla School/AMA + Bosch Arquitectos
健康和社会进步中心_Center for Health and Social Advancement/Kéré Architecture
Thread艺术家住所与文化中心_Thread Artist Residency and Cultural Center/Toshiko Mori Architect
牛背山志愿者之家_Cattle back Mountain Volunteer House/dEEP Architects
漂浮在天空中的孤儿学校_Floating in the Sky School for Orphans/Kikuma Watanabe
Trika别墅_Trika Villa/Chiangmai Life Construction
紧凑的喀斯特房屋_Compact Karst House/Dekleva Gregorič Arhitekti
P住宅_P House/Budi Pradono Architects

地方与全球_Local vs. Global/Aldo Vanini

...m and Diversity

The overwhelming force of economic and cultural globalization, as the unified model for every contemporary action, produces the disturbing idea that "global" is contemporary and modern, and "local" is ancient.

The increasing influence of the market on decisions, use of resources and, ultimately cultural choices, also now commoditized, makes decision makers at every level maintain a conformist attitude that flattens and deadens results. The local community perceives the objects most representative of this formalistic global movement as alien presences, at best useful as landmarks or visitors and tourist attractions.

On the other hand, the reaction to a planetary homologation that ignores roots and references to local cultures and traditions, is likely to lead to a vision, at best conservative, at worst pleased with nostalgic and vernacular quotations. Both scenarios cost us the richness and complexity contained in the answers that architecture could provide in every time for specific climatic and geographical conditions, even developing peculiar building techniques related to the local context.

The contrast between modernity, conceived as adherence to an international repertoire, and local culture of space and building, is a false and ill-posed problem. Regional architectural traditions, only apparently static, are, in any event, the result of an ongoing historical evolution and, for this reason, can and must create their own expression of contemporaneity.

The attention to an architecture generated by local experience, or at least from the confrontation between external experiences and genius loci, is also crucial to overcome the poverty of typological imagination that afflicts the spectacular monuments of global architecture.

New regional approaches to architecture are not nostalgic archeology of folk traditions, but contemporary spatial and anthropological conceptions closely related to the extraordinary richness of local cultures.

Just as the respect of biodiversity ensures the survival and stability of the natural environment, respect for cultural diversity guarantees the survival of an evolutionary architectural thought.

地方与全球

20世纪初就开始的真实与虚拟通信的疯狂扩张使文化、资源和本地化重新紧密结合，最终达到其巅峰状态，即当下占据主导地位的全球化。

全球化在拓宽视野和文化间互通的知识方面有着毋庸置疑的优点，但它同时也忽略了全球无数中小型文化中的重要细节。

全球化的另一个问题在于在各个方面都普遍降低了人类活动对于经济和财政的价值，使市场成为决策人的绝对参考，而决策人只考虑不偏离可靠和一般标准的决策，以此来降低失败的风险。

结果是产生了一个毫无特色的人类环境，人们无法辨别方位。另外，应用不是根据特定气候、社会和生产情况量身定制的标准化解决方案，使人们被迫采用环境影响大、可持续性低的技术性纠正措施。

对全球化技术产生的福利和商品的高期望值，也使得人们错误地认为一切事物如果想要变得时尚和现代，就必须符合国际化的标准，而个别地区的独特性只能代表过去遗留下来的陈旧思想。

同时，当地社区将全球化建筑风格公认的形式看成是外星生物，很少将其作为自身文化或具有自身特性的纪念碑的一部分，因此，就形成了一种对任何现代性表达的非理性的拒绝。

结果，认为全球性和国际性才是"现代的"，而地方和传统的则是"老旧过时的"这种模式化思想不断扩散，经常使得建筑师在地方规模项目中无法使优化和弘扬文化与传统的方法朝着更加进化的方向发展，无法整合新标准和先进的技术，一般的结果就是产生一座模仿地方性的建筑。

周边环境的现代性与本地化风格保护主义（这两种激进态度之间存在的对立性和互相不信任），使得可丰富自身能力的现代性失去了地方文化的复杂性，也使得本地化风格在未丢失与人类学和地方特色相关的基本特征的前提下，失去其遵循正常演化过程实现自身发展的能力。事实上，在这些方面，问题本身就是不正确的，或者说，问题的出现可能源于现代运动基本原则的一种退化趋势。

其实，20世纪理性主义建筑形式来自于"形式追随功能"这一理念的应用。几何形状、国际工业材料的应用、住房标准的定义都无非是转变功能以顺应都市社会和都市生活方式，而它与前几个世纪中更加注重功能性的装饰风格完全不同。

Local vs. Global

The frantic expansion of physical and virtual communications started at the very beginning of the 20th century has allowed an intense remix of cultures, resources, and localizations culminated in the now-dominant phenomenon of globalization. If globalization constitutes an indisputable asset in terms of widening vision and reciprocal knowledge among cultures, it also flattens the vital details of the planet's countless small, and less small cultures.

Another problem of globalization is the general reduction of every aspect of human activity to economic and financial values, leaving the market as the absolute reference for decision makers, whose only concern is not to depart from tried and average canons, thus not exposing themselves to the risk of failure.

The result is a flattened anthropic environment, homologated to a point where one can no longer recognize location. In addition, the use of standardized solutions, not tailored to the specific climatic, social and productive conditions, forces to introduce technological correctives generating high environmental impact and low sustainability.

The high desirability of the welfare and goods the globalization of technology produces, has also induced the false belief that everything that aspires to be modern and contemporary should respond to the canons of internationalization and that the peculiarities of individual regions represent only a legacy of the past and of conservative thought.

At the same time, local communities perceive the homologated form of a globalized architecture as alien objects, seldom recognized as part of their own culture or monument of their own identity, creating, as consequence, an irrational refusal of any expression of modernity.

As a consequence, the diffusion of the stereotype considering global and international to be "modern", and local and traditional to be "old", has often prevented architects working on a local scale from improving and developing cultures and the traditional methods towards more evolved forms, integrated with new standards and advanced technologies. The modest result is a shy vernacular fakery.

The opposition and the reciprocal distrust between these two radical positions, a contextual modernity versus vernacular conservation, impoverish both, depriving modernity of the capacity to enrich itself of the complexity of regional cultures, and vernacular of the ability to advance according to a normal evolutionary process, without losing basic characteristics related to the anthropology and to the genius loci.

Actually, in these terms, the problem is ill-posed or probably comes from a degenerative tendency in the founding principles of the Modern Movement.

In fact, the formal aspects of the rationalist architecture of the twentieth century came from a careful application of the concept "Form follows Function". The geometric shapes, the use of materials of the international industry, and the definition of housing standards, were nothing more than the transformation of functions congenial to metropolitan society and the

那些所谓的功能性解决方案时常会演变成风格主义,而得益于资源和技术的可用性,它们才能够克服环境需求和地方特性之间的矛盾。

尽管时常有人对某些嵌入结构的环境的可持续性提出异议,但随着archistars系统模型及其少数拙劣模仿者的出现,这一形式主义的内涵还是得到了巩固。

相反,通过更加严格地应用现代运动的原则,建筑师重新考虑了应用本地化应对措施,而不是本地化形式。它们不是过去封闭社区的怀旧风格纪念物,如成为种族和民间传说的一部分,而是成为对随时间推移不断发展的当地状况所提出挑战的高效功能性回应。

然而,基本功能不仅仅能为节能和气候可持续性问题提供对策。建筑将它们自己确立为一个社区的关注点,具有最深层和最本真的纪念意义。通过使用原型来巩固构建的方式也清楚地表明,这些方式远非构成仿制复本的静态模块,它们代表的是建筑师用以明确表达并且顺应当代建筑思想的实体材料。针对环境关系展开设计是一种一般性方法,它并不一定要与当地建筑师相关,也可以与该地区的特征相关。

当地建筑知识提供的机遇远不止体现在建筑的技术方面。与基于国际标准的类型匮乏的项目相比,不仅就功能范畴,甚至就问题的解决方案而论,对建筑需求的地方性回应都能提供更多的类型和解决方案。它为当今学科中最薄弱领域之一的类型学研究奠定了基础。

对于建筑文化环境,目前同样需要我们关注生物多样性。物种多样化的复杂性和丰富性能确保实现更加坚固和具有弹性的环境系统,同样,建筑风格多样化能确保在功能的可持续性、一致性和自我认知方面实现更加坚固和具有弹性的地方社区。这些都是促进小地方社区发展,防止其发展成城市群的基本方法,因为城市群缺乏地方特色,生活质量不高。

为适应地方和文化特征,本书的案例将展示它们如何摆脱一成不变的国际化模式,应用社区知识、传统材料和当地工匠,促进社区老建筑和新建筑之间的关系。

位于孟加拉拉贾尔哈特的帕尼社区中心由荷兰建筑师事务所SchilderScholte设计,我们可以从中发现以上所有的特征,从当地材料的精心利用,到对这一分布于广阔森林中的社区生活方式的解读,我们可以明智地将其重新看作是改善和更新当代建筑语言的一个起点。几乎实现零公里建筑,所有的建筑材料都来自于距离建筑场地25km的

metropolitan lifestyle, as opposed to the more semantic and functional decorative style of the previous centuries.
Those that were functional solutions have often evolved into stylistic mannerisms that, thanks to the availability of resources and technology, could overcome inconsistencies in environmental needs and regional characters.
With the emergence of models of the archistars system and its minor epigones, this formalistic connotation consolidates, despite frequent questionable claims of environmental sustainability of some interventions.
A more stringent application of the principles of the Modern Movement leads, on the contrary, to a reconsideration of local measures, not necessarily local forms. They become not as a nostalgic relic of the past of isolated communities, such as ethnic and folkloristic repertoire, but functional and efficient responses to the challenges posed by local conditions, developed over a long time.
The fundamental functions, however, provide more than answers to problems of energy efficiency and climate sustainability. The buildings establish themselves as points of recognition of a community, in the deepest and most authentic meaning of the monument. Consolidation of building modes by archetypes makes it clear that, far from constituting static models for an imitative replication, these modes represent the solid material on which to articulate and adapt a contemporary architectural thought. A relationship toward the context can be a general approach, not necessarily connected to local architects, but connected to the character of the location.
The opportunities offered by local architectural knowledge go far beyond the mere technical aspects of building. Compared to the impoverishment of the typological range of projects based on international standards, not only in reference to functional categories, but even to their solutions to problems, the regional responses to the request for architecture provide a much more varied range of types and solutions. This establishes a fundamental ground for typological research, which today constitutes one of the weakest areas of the discipline. To the context of architectural culture, the same attention currently addressed to biodiversity is necessary. In the same way that the complexity and richness of species diversity is a guarantee of more solid and resilient environmental systems, architectural diversity ensures more solid and resilient local communities, both in terms of functional sustainability, identity, and self-recognition. These are fundamental to the strengthening of settlements threatened by the centripetal trend towards urban mega – agglomerations, lacking genius loci and therefore quality of life.
In adapting to the specificity of the places and cultures, the examples show how it is possible to deviate from the homologation of international models, using the knowledge of the community, traditional materials and the local craftsmen, facilitating the relations between the traditional customs of the community and new interventions.
In the Pani Community Center in Rajarhat, Bangladesh, commissioned to the Dutch architects SchilderScholte, we find all the above characters, from the careful use of local materials

半径范围内，形成一个独立的U形竹子结构和可回收再利用波纹板外壳下的环保型建筑体量。甚至为数不多的用来强调的颜色都来自于当地有关驱赶昆虫的知识。然而，在未明显分离内外部的情况下，它更多的是体现与传统空间理念密切相关的类型学解决方案。而结果是产生一栋不采用预制模块，但尊重当地生活习惯的建筑：一个在形式和功能方面都得到解决的实验室。

有效控制使用极少的构成组件，使得由AMA建筑事务所和Bosch建筑事务所联合设计的秘鲁丘基班比亚学校看起来焕然一新。安装、形状和材料应用都极其简单，是利用应对气候舒适度的被动解决方案。按照设计方案，建筑师不单纯地将学校看作是一栋教学楼，而是将它作为社区的中心，以此避免应用那些可能威胁到简单个别群体的不适用的模块。

通过利用貌似小屋屋顶的大面积斜坡表面来收集农用灌溉雨水作为项目的重点，位于塞内加尔Sinthian村的Thread艺术家住所与文化中心项目，解决了非洲大陆的传统需求之一——水源稀缺问题。而这一对于西方世界目前仍陌生的问题，成为纽约建筑师森俊子项目设计的第一主题，他以一种现代化的方式解决了这一问题，但却仅使用了当地的材料。他定义了当地文化的典型空间，该空间几乎整体都由带顶的开放空间组成，从市场到集会区，它在社会文化的任何时刻都发挥功能。该项目由阿尔伯斯基金会推广，体验"从零开始"，它也正是艺术家和创始人约瑟夫和安妮·阿尔伯斯在该起点下找寻到完美应用的基本理念。

文化视野上的重大改变推动了形式与功能方面的复杂结构试验，这种复杂结构结合了宗教观念、教育以及建筑师为泰国Sangkhlaburi村孤儿构思的设计想法。菊间渡边将对"飞船"的幻想转变成了一所梦想学校，通过一座"发射台"收集地球能量，它的建造完全跳出了固有的条条框框，形成一种以传统原型和青年顾客的灵感启发为主要特点的全新类型学建筑。

位于中国蒲麦地村的牛背山志愿者之家，是一家为致力于维护村庄的青年志愿者服务的旅社，虽然在功能实现方面缺乏创造性，但它却在控制和提升传统屋顶设计方面探求了数字技术所能提供的设计机遇，成为一座强大的地标性建筑，以此充分证明了传统和创新之间、人工和自然形态之间可能存在的协同作用。

在印度尼西亚沙拉笛加的P住宅项目中，只有严格按照其环境状况开展设计才能获得明亮并且具有低环境影响的解决方案。Budi Pradono建筑师事务所实施了一个房屋设计计划，在这个项目中甚至使

to an interpretation of the lifestyle of a community scattered over a vast forest, intelligently reconsidered as a starting point to improve and upgrade to a contemporary language. In a sort of zero-kilometer architecture, all the building materials come from within a 25km radius of the site, resulting in environmentally friendly volumes placed under the cover of an independent U-shaped structure of bamboo and recycled corrugated sheet. Even the few color accents come from local knowledge in terms of repulsion of insects. However, more relevant is the typological solution strictly related to a traditional concept of the space without a clear separation between internal and external fruition. The result is a building not attributable to pre-established models, yet respectful of local living habits: a laboratory of formal and functional solutions.
A wise control of very few constituent elements keeps AMA-Afonso Maccaglia Architecture+Bosch Arquitectos' design for the Chuquibambilla School, Peru, from looking shabby. The extreme simplicity of installation, shapes and materials, the use of passive solutions addressed to the climate comfort, complies with a program that views the school as much more than a building for education, but the focal point of the community, avoiding foreign models that would intimidate a very simple and isolated population.
Focused on collecting rainwater for the agriculture by the large sloping surfaces of the hut like roof, the project for the Thread Artist Residency and Cultural Center for the village of Sinthian in Senegal, addresses one of the traditional demands of the African continent, the scarcity of water. A problem now unknown in the Western world became the dominant theme for the New York architect Toshiko Mori, who worked it out in a contemporary way, but through the exclusive use of local materials. He defined a typical place of the regional culture, almost entirely made up of covered but open spaces, functional to all the moments of that social culture, from the market to the assembly. The project, promoted by the Albers Foundation, experiments with "starting at zero", a basic concept of the artists and founders Josef and Anni Albers that finds perfect application in this context.
A drastic change of cultural horizon leads to experimenting with a complex configuration of forms and functions combining religious sense, education, and the ideas solicited by the architect to the orphans of the Thai village of Sangkhlaburi. Kikuma Watanabe transformed the fantasy of a "flying boat" into a dream school that collects the earth energy through a "launching pad". It drew, completely out of the box, a totally new typology that feeds on traditional archetypes and the inspirations of the young guests.
Less innovative in implementation of functions, the Cattle Back Mountain Volunteer House in the Chinese village of Pumaidi, a hostel for young volunteers devoted to village maintenance, explores the chances offered by digital techniques to control and upgrade the traditional roof design that becomes a strong landmark, in an evident demonstration of the possible synergy between tradition and innovation, and between artificial and natural shapes.
Only with close adherence to the conditions of the context is

用了当地材料和技术,为一个大家庭创造了一个明亮的生活空间,该项目尤为注重实现森林和周围环境之间最大限度的感知交流。这样的经历使得当地工作者接触到了新的设计观点,确保了对古老建筑知识的延续。

Chiangmai Life Construction事务所设计的位于泰国清迈的Trika别墅突破了居住空间必须以使居住者与其周边环境隔离的封闭箱体状外形存在的设计理念。该项目致力于实现极高的目标,借助于竹子和土壤重新将居住空间诠释为顶部开放的围护结构,以实现更好的通风效果并保持与周围森林的亲密接触。由柏油薄片和劈裂的竹梁组成的薄"伞状结构",令人们想起传统建筑的样式,它能保护建筑免受恶劣天气的影响。

德国导演Christoph Schlingensief和多民族设计团队Kéré建筑事务所共同在布基纳法索平原中心地带创建了一个项目,以建造一个能够向当代世界传播传统文化的艺术精品中心。它是目前构思出的整个推理的缩影。而歌剧村几乎可以说是一个史无前例的设计方案,该项目的施工完全尊重地方认同。来自不同领域的学生和艺术家可以聚集在歌剧和艺术表演中心。项目伊始设计者就利用了当地的人类学特征,包括使用当地材料,但最终实现的却是现代感十足的外形设计。其中

最早完工的建筑物之一为健康和社会进步中心。尽管医疗服务活动在第一世界国家被严格地标准化了,但对于距离西方世界体制还很遥远的国家来说,对独特建筑形式的想象更符合当地社区的生活方式和期望值。

由Dekleva Gregorič建筑事务所开展的位于斯洛文尼亚喀斯特村的房屋设计项目,也将对传统建筑的分析和当地材料石灰岩的使用作为项目的起点。这种石料不仅在场地上随处可得,在历史上,这种材料甚至曾在威尼斯市的起源地被提及,它是在威尼斯支柱引起森林砍伐时为人所知的。事实上,该项目没有参照地方建筑风格,而是参照最深层的建筑原型,形成一个典型的设计方案和一个近乎于孩子气的房屋设计想法。然而,既巧妙又简单的设计解决方案将这一基本理念转变成一个连接地方传统和国际现代性的一座典雅的桥梁。

虽然我们这里讨论的项目本身并不能代表建筑学中出现的所有问题的整体解决方案,但它们体现了基于关注环境和地方性知识的解决方法,而这种方法超出了个案,它适用于世界大多数地区的情况——从最与世隔绝的自然环境到大型的城市周边环境。

possible to obtain the lightness and the low environmental impact of the P House in Salatiga, Indonesia. Budi Pradono Architects multiplied an archetypal house scheme and, even in this case, using local materials and techniques, have created a bright living space for an extended family, with great attention to the maximum perceptive exchange with the forest and surroundings. Such experiences allow native workers to deal with new design perspectives, ensuring continuity with an ancient building knowledge.
Chiangmai Life Construction, in the Trika Villa in Chiang Mai, Thailand, goes beyond the concept of habitable space as sealed boxes that isolate the people from their surrounding environment. While addressing a very high target, the project resorts to bamboo and earth to reinterpret living space as an enclosure open at its top for better ventilation and intimate contact with the forest. Only a thin "umbrella" of tar sheet and split bamboo beams, reminiscent of traditional forms, protects it from the elements.
German director Christoph Schlingensief and the multi-ethnic team Kéré Architecture, working in the heart of the plains of Burkina Faso, created a center of artistic elaboration capable of projecting traditional culture to the contemporary world. It epitomizes the whole reasoning developed so far. Opera Village is a virtually unprecedented program, carried out in full respect for local identity, which accommodates students and artists of various disciplines gathering around a central core for the opera and the performing arts. They began by using the anthropological characteristics of the region, including locally available materials, but with definitely contemporary formal results. One of the first buildings of the structure they completed is the Center for Health and Social Advancement. Even an activity so rigidly standardized in first-world societies as health service can be, for countries still far from Western organization, it is a chance to imagine unique building types better adapted to the lifestyle and expectations of local communities.
Also the project by Dekleva Gregorič Arhitekti for the small house in the Slovenian village of Karst, starts from the analysis of traditional architectures and from a material, limestone. This stone is not only readily available on site, but even it historically refers at the origin of the city of Venice, when it was brought to light by the deforestation needed for the Venetian stilts. In fact, rather than to regional architectural references, the project refers to the deepest architectural archetypes, to a prototypical scheme and to an almost childlike idea of home. However, the introduction of sophisticated and minimalistic solutions transforms this basic concept into an elegant bridge between regional tradition and international contemporaneity. Although the projects discussed here cannot represent, by themselves, a comprehensive solution for all the problems posed to architecture, they embody an approach based on an attention to context and to local knowledge, which goes beyond the individual case but applies to most planetary situations, paradoxically, from the most isolated and natural locations to the largest urban peripheries. Aldo Vanini

帕尼社区中心
SchilderScholte Architects

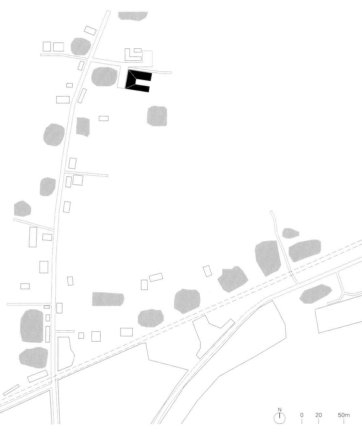

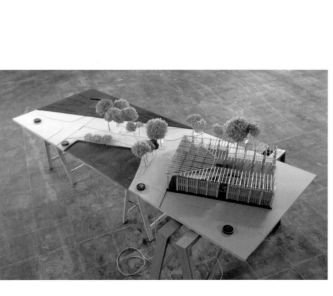

荷兰帕尼基金会委托SchilderScholte建筑师事务所来设计位于孟国拉北部的拉贾尔哈特小镇的教育大楼。基金会的主要目的是培养就业前景不好的当地居民的专业技能，使他们可以自主经营。

建筑师夫妇接受了这个源于意识形态和知识共享的公益项目。在设计过程中，建筑师的设计重点主要是当地现有的材料和气候条件。设计出发点是实现建筑材料和技术均来源于场地25km半径范围内。这一努力意在鼓励当地建筑商了解可持续发展的基本原则。实际上，在这座建筑的建设过程中以及为此而需要的其他过程中使用的电力或化石燃料接近于零。

建筑平面（24m×32m）是东西向的，包括位于大型竹屋顶结构下

的三个体量。两条视线可以穿过建筑的四个方向。南侧可以看到带有卫生间的教室,北侧有工作室及仓库。出于生物气候学的考虑,这样的朝向可以增强自然交叉通风。建筑体量之上的漏斗形屋顶向上抬升了一段距离,因此显著减少了室内热量的积聚,还可以将雨水收集进庭院。周围的植被和附近的池塘能提供自然通风,进一步降低了温度。

一些仿生学的元素在这里得到了检验!

砖建造的体量全部抹灰,部分涂漆。室内墙面的颜色为淡蓝色,这是苍蝇会避开的色调。教室窗户的侧面都漆成黄色,某些昆虫不喜欢这种色调。黄色也代表了芥菜的花朵,这一作物从十二月到一月开遍孟加拉大部分地区。灰色和黑色又代表了孟加拉土地降雨之前和之后的颜色。虽然竹子被视为该地区的劣质材料,但是建筑师依然选用竹子建造整个屋顶结构。甚至连工作室的墙壁和落地门也用竹子包覆。这样的设计参考了当地生产的竹制自行车框架。

U形的砖柱支撑建筑的南立面,从而创造出一排小型垂直窗户。最终这一侧立面可以理解为一面石墙,这极大地节约了建筑成本、时间和劳动力。窗户尺寸的选择非常谨慎,这样的设计将直接照射到教室的阳光减到最少,同时仍能提供最佳的日光照明。应用当地砖比较便宜,而竹子和薄混凝土楼板的结合使用减少了结构木材的使用,从而最大限度地减少了对孟加拉地区木材短缺的影响,降低了日后的维护成本。

Pani Community Center

The Dutch foundation Pani commissioned SchilderScholte Architects to design an educational building in the north Bengal town of Rajarhat. The main objective of the foundation is to train prospectless local people professional skills and become autonomous.

The couple architects embraced this pro bono assignment coming from ideological motives and knowledge sharing. During the design process attention was mainly focused on locally available materials and weather conditions. The starting point was to realize a building using materials and skills from within a 25km radius from the site. The drive was to encourage local builders to become aware on the basic principles of sustainability. In effect close to zero electricity or fossil fuels were used during construction and other necessities required for erecting this building.

The floor plan (24 x 32m) is east-west oriented and consists of three volumes under a large bamboo roof structure. Two sight lines traverse the building in all four directions. On the south side we find the classrooms with lavatories and on the north side the workshop and store. From a bioclimatic point of view

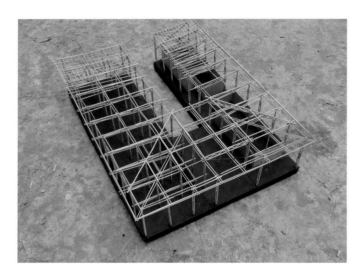

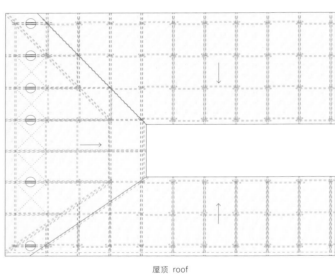

屋顶 roof

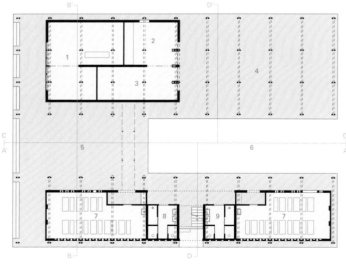

1.仓库 2.工作室 3.储藏室 4.工作间 5.广场
6.庭院 7.教室 8.女士卫生间 9.男士卫生间
1. store 2. workshop 3. storage 4. workplace 5. plaza
6. courtyard 7. classroom 8. ladies lavatory 9. gents lavatory
一层 first floor

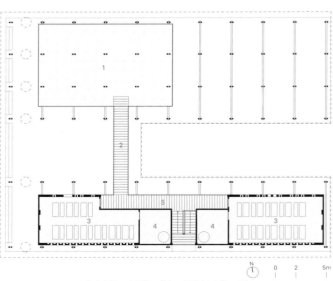

1.集会大厅 2.步行桥 3.教室 4.储藏室&水箱 5.平台
1. assembly floor 2. footbridge 3. classroom
4. storage & water tank 5. landing
二层 second floor

1 安排基础，埋入柱基
1. lay out foundation blocks embed column plinth

2 安装边缘框架
2. install edge formwork

3 安装柱子
3. erect columns

4 安装天花板格栅
4. install ceiling joists

5 安装斜桁架与梁
5. install diagonal truss and beam

6 交替在屋顶椽条内安装脊梁，注意：倾斜安装
6. install ridge beams in roof rafters alternately mind: sloped install

7 安装天花板椽条
7. install ceiling rafters

8 安装风力支撑，水平和垂直安装
8. install wind bracing, both horizontal and vertical

9 安装凹口椽条，安装在吊顶和屋顶上
9. install notch rafters, both dropped ceiling and roof

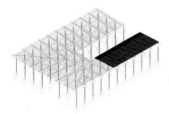

10 安装波纹板与天花板垫
10. install corrugated panels and ceiling mats

11 第二阶段，体量
11. phase 2, volumes

项目名称：Pani Community Center
地点：52 Hospital Road, Makurtary, 5610 Rajarhat, Bangladesh
建筑师：SchilderScholte Architects
项目团队：Gerrit Schilder, Hill Scholte
合作者：Dick van Gemerden _ pt-structural
用地面积：816m² / 建筑面积：658m² / 有效楼层面积：910m²
设计时间：2012 / 施工时间：2013—2014
摄影师：courtesy of the architect

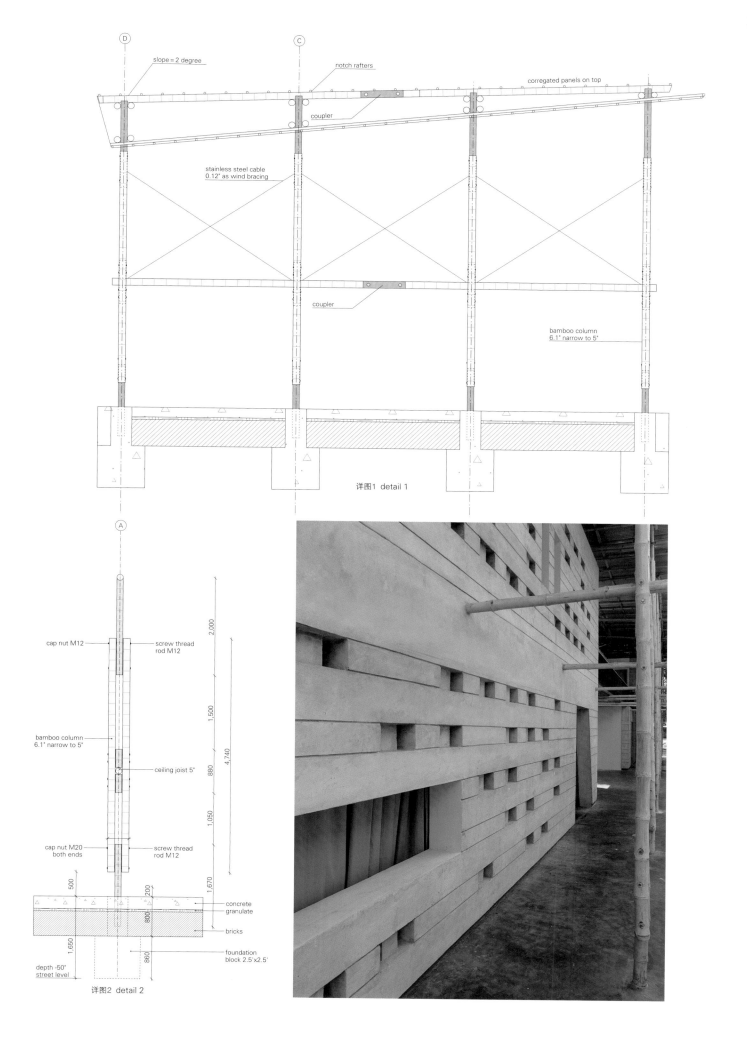

详图1 detail 1

详图2 detail 2

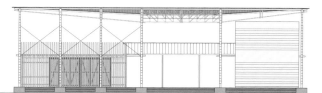
西立面 west elevation

北立面 north elevation

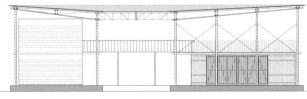
东立面 east elevation

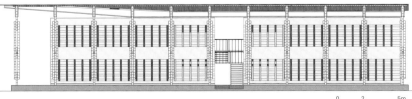
南立面 south elevation

0 2 5m

the orientation allows to emphasize the natural cross ventilation. The lifting of the funnel-shaped roof high above the volumes has achieved a considerable reduction of heat build up within the spaces and collects rainwater into the courtyard. Further cooling is provided by surrounding vegetation and the use of nearby ponds for natural draft.
Some biomimicry elements are put on the test here! The brick-built volumes are all plastered and partly painted. The interior walls are coloured in light blue, a hue that flies shun. The splays of the classroom windows are painted yellow, a hue that specific insects dislike. Yellow also refers to the flowers of the mustard plant, a crop that colors large parts of the country from December to January. Grey and black in turn refer to the color of the Bengal earth before and after rainfall. Although bamboo is seen as an inferior material in the region, the architects have chosen to make the whole roof construction out of it. Even the walls and French doors of the workshop are cladded with it! Thus this design is a reference to the bamboo bicycle frames that are made here.

U-shaped brick columns support the south facade of the building, thus creating a row of small vertical windows. Ultimately, this can be understood as a single stone wall, a great saving on construction costs, time and labor. The dimensions are chosen with great care, in such a way that direct sunlight into the classrooms is minimized while still providing in optimal daylight illumination. This application of local bricks is less expensive and reduces the use of construction wood by combining bamboo with thin concrete floors. This minimizes the shortage of wood in Bangladesh as well future maintenance costs.

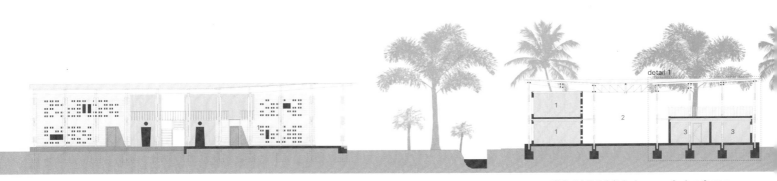

1.广场 1. plaza
A-A' 剖面图 section A-A'

1.教室 2.广场 3.仓库 1. classroom 2. plaza 3. store
B-B' 剖面图 section B-B'

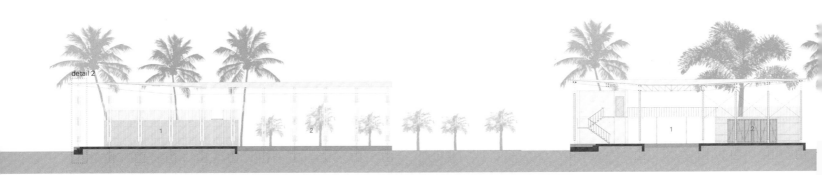

1.广场 2.工作间 1. plaza 2. workplace
C-C' 剖面图 section C-C'

1.庭院 2.工作间 1. courtyard 2. workspace
D-D' 剖面图 section D-D'

丘基班比亚的当地社区位于秘鲁高高的丛林之中，是Pangoa东部地区最重要的咖啡产区，也是该地区的文化中心。其儿童人口约为250人，学习条件极差。他们中有些人不得不长途跋涉去接受教育，甚至根本没有受教育的机会。由于是印第安本土的社区，人们根据自己的文化和习俗生活。他们从事农业、狩猎和捕鱼。社区没有电、自来水和污水处理系统。

在丘基班比亚，学校是具有强烈的社会责任的项目，社区居民都要参与到项目中来，场地的实际需求和不足也经过了仔细研究。学校不仅要在上课期间提供一个教育场所，还要成为一个充满生命力的、服务于整个社区发展和交流的地方，让家长、学生和老师可以在一起学习和娱乐。该项目设置有三个教学模块和一个居住模块，它们都围绕着中央庭院设置。这些模块包括教室、教师办公室和行政管理区、多功能教室（图书馆，工作室等）、电脑室和学生的宿舍。

该项目有一个大型的户外功能区，是一个由不同规模的有顶庭院组成的系统，可举办各种活动，使学生们多多接触自然和他们的传统。这些活动包括户外课、艺术品和黏土制品制作、手工艺品制作、农艺学、畜牧、作物学习等。这些空间通过有遮挡的道路连接，这条道路被设计为利用率较高的会面场所，并且成为项目的扩展区域。建筑内外之间的界线模糊，创造了与周围环境相连的公共空间。

建筑设计以及抗震结构设计结合了本土和现代材料，利用当地资源引入现代建筑系统。建筑师雇用当地劳工，通过现场施工传授知识和经验。设计的目标是给居民带来一种归属感，激励他们在周围环境中继续建造此类建筑。建筑师通过采用被动式系统实现了气候舒适性，并特别关注了光照控制、通风和自然采光，最大限度地降低了能源需求。电脑室由太阳能电池板供电。灰水经过处理后重新用于绿地植物灌溉。

由于文脉和当地特殊的情况，该项目在社区的参与下发展起来。由学生、教师和志愿者共同组成的工作室帮助这个有多种文化的社区实现了思想的交流。

丘基班比亚学校

AMA-Afonso Maccaglia Architecture + Bosch Arquitectos

西立面 west elevation

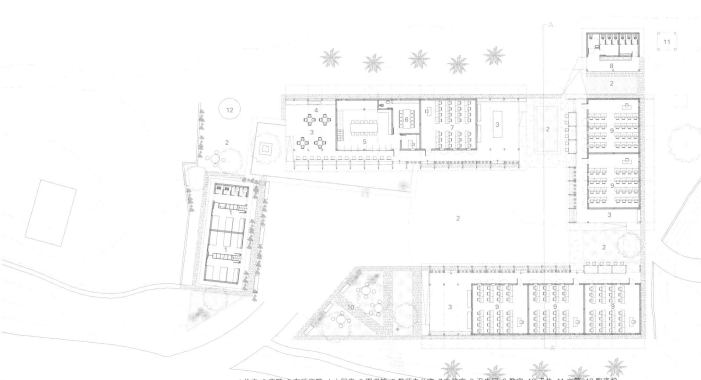

1.住宅 2.庭院 3.有顶庭院 4.小厨房 5.图书馆 6.教师办公室 7.电脑室 8.卫生间 9.教室 10.天井 11.水箱 12.陶瓷炉
1. residence 2. courtyard 3. covered courtyard 4. kitchenette 5. library 6. teacher room
7. computer room 8. toilet 9. classroom 10. patio 11. water tank 12. ceramic oven

一层 first floor

项目名称：Chuquibambilla School / 地点：Comunidad Nativa de Chuquibambilla, Satipo, Peru
建筑师：Paulo Afonso, Marta Maccaglia, Ignacio Bosch, Borja Bosch
结构工程师：Manuel Cardenas Aspajo / 施工：Angel Javier Garcia Paucar_JIC S.A., Local community
木匠：Elias Josè Martinez Ramos_Carpinteria Martinez, Maximo Ñhaui Centeno_Carpinteria ÑHAUI
资金提供：Costa Foundation, Volcafe Foundation, ED&F MAN ChariCo
合作者：Procesadora del Sur S.A.
用地面积：4,000m² / 建筑面积：985m² / 造价：USD 185,000
设计时间：2012.10—2013.3 / 施工时间：2013.4—2013.12 / 摄影师：courtesy of the architect

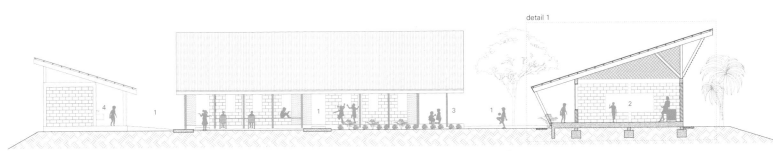

1.庭院 2.教室 3.有顶庭院 4.卫生间 1. courtyard 2. classroom 3. covered courtyard 4. toilet
A-A' 剖面图 section A-A'

Chuquibambilla School

The native community of Chuquibambilla, located in the Peruvian high jungle, is the most important coffee zone of the eastern district of Pangoa and is the cultural center of the region. The children population is about 250, who are studying in very poor conditions. Some of them have to travel long distances or do not even have access to education. As it is an Indian native community, people live according to their culture and customs. They are engaged in agriculture, hunting and fishing. The community has no electricity, running water or sewage system.

The school in Chuquibambilla is a project with a strong social burden, where the community comes to be part of the process, and where the real needs and deficiencies of the site are researched. More than just a place of education during school hours, the school seeks to be a place of development and exchange for the whole community, always alive, where parents, students and teachers can meet for study and recreation. The program is set with three school modules and a residential module arranged around a central courtyard. These modules contain classrooms, teachers and administration area, a multifunctional classroom (library, workshops, etc.), a computer room, and dorm for students.

The project has a large outdoor program through a system of covered courtyards at various scales, for activities that connect students to nature and their traditions: outdoor classes, art and clay workshops, crafts, agronomy, animal husbandry, crop, etc. The spaces are connected by a shaded path that is assumed as an effective meeting place, becoming an extension of the program. A building in which the boundaries between interior and exterior fade to create public space connected to its environment.

Along with a seismic structural design, the building design combines vernacular and modern materials, introducing modern building systems using local resources. The inclusion of local labor allows the transfer of knowledge through on-site experience. The ambition is to create a sense of belonging among residents and to inspire a process of continuing work in its environment. Climate comfort is achieved through the use of passive systems, with particular attention to sunlight control, ventilation and natural lighting, reducing energy requirements to a minimum. The computer room is powered by solar panels. Grey water is treated and re-used for irrigation of green areas.

The project, due to its contextual and local specificities, has been developed through a participatory process with the community. Workshops with students, teachers, and volunteers have been instrumental to bring this culturally different community to an exchange of ideas.

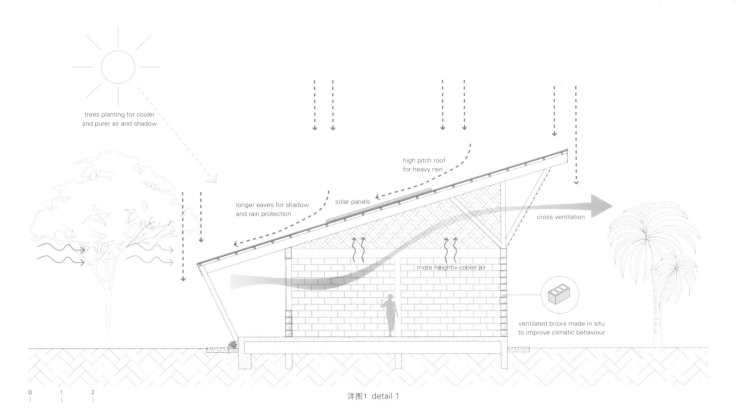

详图1 detail 1

健康和社会进步中心
Kéré Architecture

地方主义与全球多样性 Regionalism and Global Diversity

虽然布基纳法索是世界上最贫穷的国家之一，但也是一个具有强烈民族自豪感的国家。

歌剧村项目是与德国戏剧和电影导演Christoph Schlingenseif合作开发的，他的设计目标是帮助塑造和唤醒社区内的文化认同，并将其作为积极发展的战略。

该项目的最初理念是通过这一世界一流的表演中心的建设，让人们关注布基纳法索，并使其成为非洲电影和戏剧的中心。尽管大量洪水冲毁了项目现场及周围的村庄，但决策者还决定从歌剧院的建设中抽出部分资源和资金，以便将住宅、教育设施和医疗设施纳入规划。为了实现多种功能与用途，建筑师开发了适应性强的结构模块，其中集成了被动式通风、太阳能使用以及水收集和水管理等功能。

这些模块由当地黏土、木头和红土石头制成，最大限度地利用了可以在现场获取的材料，将对生态和成本的影响降到了最低。

健康和社会进步中心，也被称为CSPS，致力于为当地居民提供基本的健康和医疗资源。该中心由围绕一个中心接待区设置的三个单元组成，包括口腔科、妇产科和普通药品区。

该设施还包括检查室、住院病房和员工办公室。建筑师特别考虑到游客和患者家属会面与等待的需要，设置了若干个带遮蔽的庭院。

有趣的开窗设计出现在不同的有利位置，方便站立、坐着或者是卧床不起的人们（包括儿童）向外观看。这些窗户像相框一样，每个窗户都能给人独特的视角来欣赏景观中独特的部分。

与歌剧村的材料美学与生态相一致，当地的黏土和红土石头被用在双层墙体的建造中，可提供额外的防雨保护。当地的桉木因为加速了荒漠化而被看作是环境公害，在该项目中被用来建造中心的吊顶和有顶行人通道。

Center for Health and Social Advancement

Although Burkina Faso is one of the poorest countries in the world, it is also a nation with a strong sense of national pride. The Opera Village project was developed in collaboration with the German theater and film director Christoph Schlingenseif, whose aim was to help shape and awaken a cultural identity within the community as a strategy for positive development.

The initial concept for the project was to draw attention to Burkina Faso as a center of African film and theater with the construction of a world-class performance center. After massive flooding damaged the project site and surrounding villages though, the decision was made to pool resources and funds from the opera to introduce residential, educational, and medical amenities into the plan. In order to support this wide range of function and use, an adaptable structural

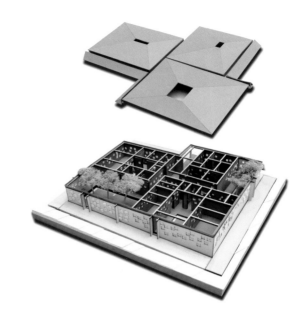

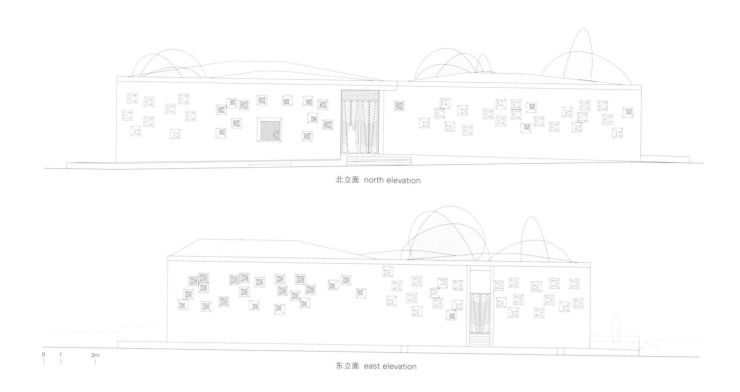

北立面 north elevation

东立面 east elevation

module was developed with integrated passive ventilation, solar energy use, and water collection and management. Made with local clay, wood, and laterite stone, these modules minimize ecologic and cost impact by maximizing the use of materials widely available on site.

Center for Health and Social Advancement, otherwise known as the CSPS, is geared towards providing basic health and medical resources for the local population. The center consists of three units organized around a central reception area: dentistry, gynecology and obstetrics, and general medicine. The facility is rendered with examination rooms, inpatient wards, and staff offices. Special consideration is taken for visitors and family of patients with several shaded courtyards for gathering and waiting.

The playful fenestration design emerged from the varying vantage points of standing, seated, or bedridden individuals including children. The windows are composed like picture frames, with each individual view focused on a unique part of the landscape.

In keeping with the material aesthetic and ecology of the Opera Village, local clay and laterite stone were used in the double-envelope construction of the walls for extra rain protection. Locally available eucalyptus wood, seen as an environmental nuisance as it contributes to desertification, is used to line the suspended ceilings and covered walkways of the center.

气候示意图 climate diagram

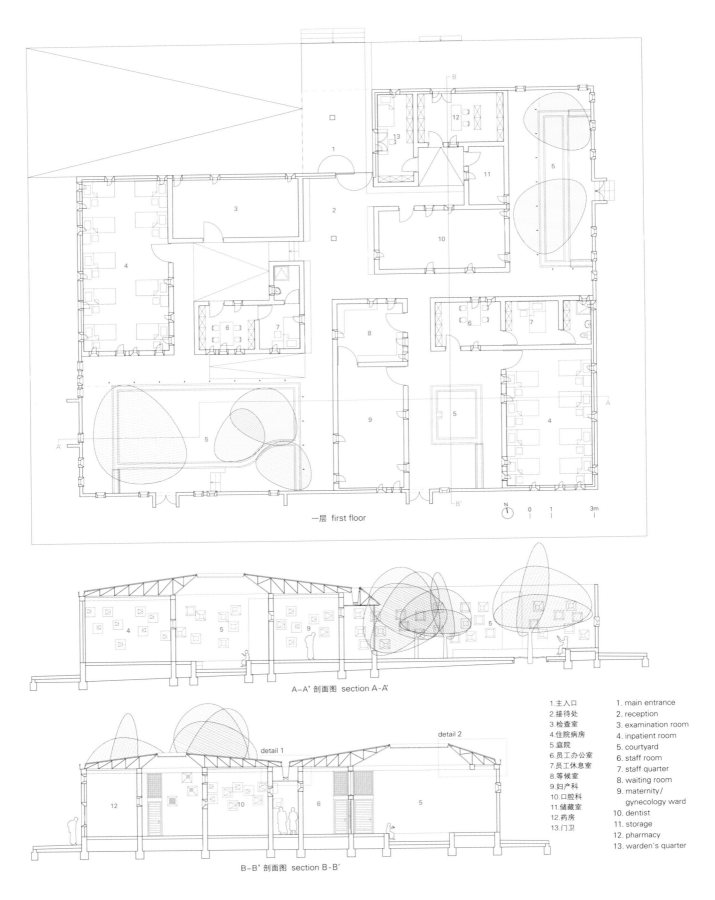

一层 first floor

A-A' 剖面图 section A-A'

B-B' 剖面图 section B-B'

1. 主入口	1. main entrance
2. 接待处	2. reception
3. 检查室	3. examination room
4. 住院病房	4. inpatient room
5. 庭院	5. courtyard
6. 员工办公室	6. staff room
7. 员工休息室	7. staff quarter
8. 等候室	8. waiting room
9. 妇产科	9. maternity/gynecology ward
10. 口腔科	10. dentist
11. 储藏室	11. storage
12. 药房	12. pharmacy
13. 门卫	13. warden's quarter

项目名称：Center for Health and Social Advancement / 地点：Laongo, Burkina Faso / 建筑师：Kéré Architecture
设计团队：Diébédo Francis Kéré, Ines Bergdolt, Emmanuel Dorsaz, Jin Gul David Jun, Pedro Montero Gonzales, Dominique Mayer
结构工程师：Grünhelme e.V. / 施工监理：Kéré Architecture / 植被工程师：Kéré Architecture, Grünhelme e.V.
景观设计：Kéré Architecture / 客户：Festspielhaus Afrika gGmbH / 用地面积：1,4230m² / 建筑面积：1,340m² / 有效楼层面积：1,340m²
材料：exterior walls _ cement hollow bricks, loam plaster, interior walls _ BTC compressed earth blocks, roof _ metal, Facade _ concrete frames for window openings, terraces _ laterite stone / 设计时间：2012 / 施工时间：2013 / 竣工时间：2014.11
摄影师：courtesy of the architect-p.49, p.50~51, p.52, p.53top / ©Erik-Jan Ouwerkerk(courtesy of the architect) -p.46~47, p.53bottom, p.55

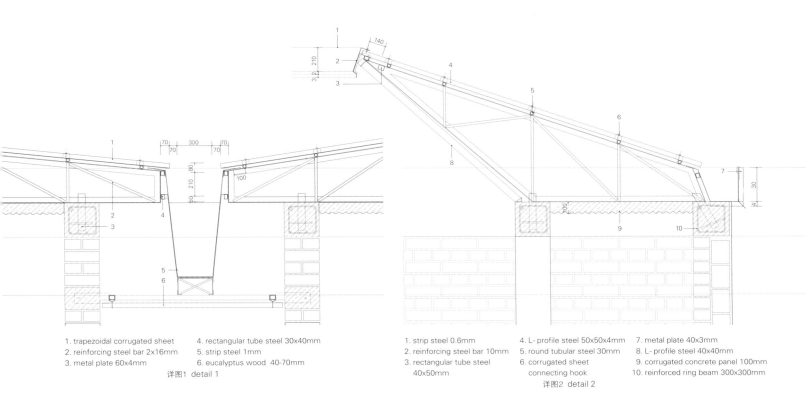

1. trapezoidal corrugated sheet
2. reinforcing steel bar 2x16mm
3. metal plate 60x4mm
4. rectangular tube steel 30x40mm
5. strip steel 1mm
6. eucalyptus wood 40-70mm

详图1 detail 1

1. strip steel 0.6mm
2. reinforcing steel bar 10mm
3. rectangular tube steel 40x50mm
4. L- profile steel 50x50x4mm
5. round tubular steel 30mm
6. corrugated sheet connecting hook
7. metal plate 40x3mm
8. L- profile steel 40x40mm
9. corrugated concrete panel 100mm
10. reinforced ring beam 300x300mm

详图2 detail 2

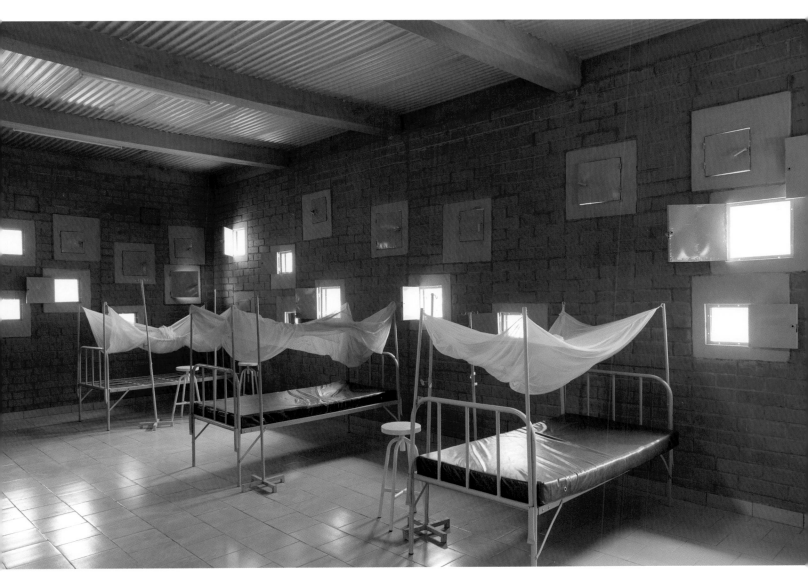

地方性与全球多样性 Regionalism and Global Diversity

Thread艺术家住所与文化中心
Toshiko Mori Architect

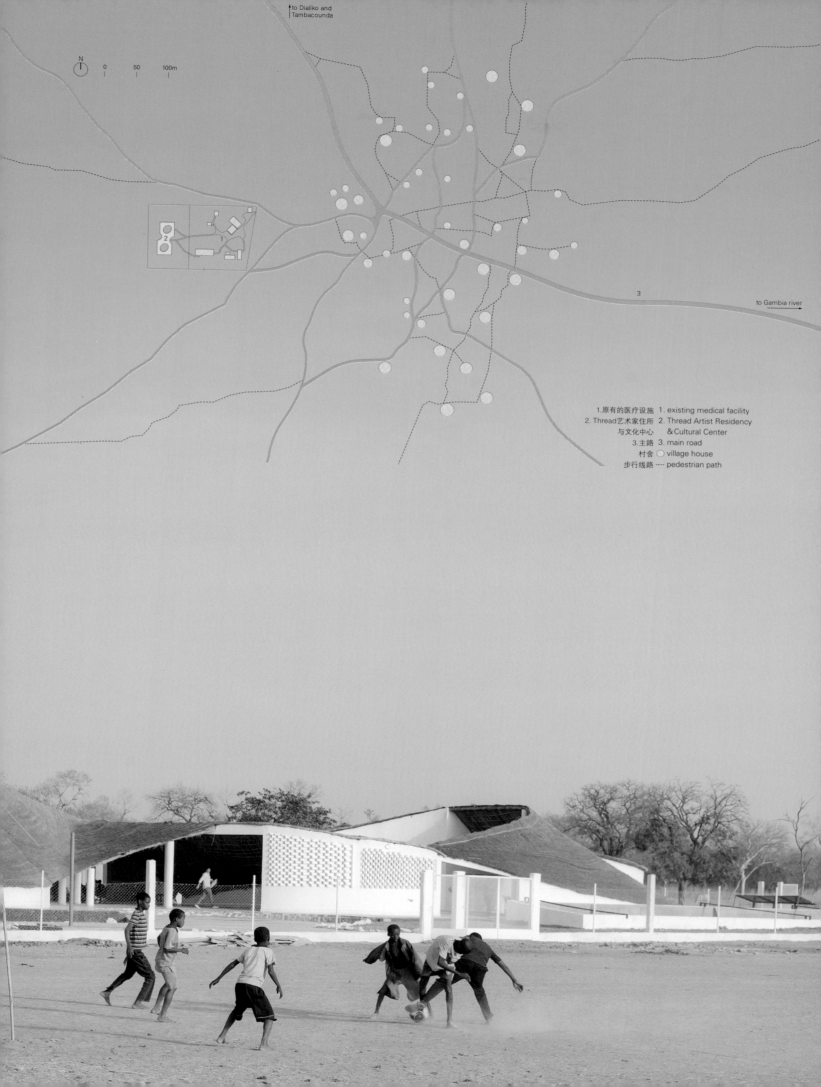

1. 原有的医疗设施	1. existing medical facility
2. Thread艺术家住所与文化中心	2. Thread Artist Residency & Cultural Center
3. 主路	3. main road
村舍 ○	village house
步行线路 ----	pedestrian path

to Dialiko and Tambacounda

to Gambia river

塞内加尔东南部的Sinthian乡村，将是一个激动人心的新文化中心的所在。该文化中心由康涅狄格州的约瑟夫和安妮·阿尔伯斯基金会构思和资助，与Sinthian当地一位领导合作开发。Thread中心提供了艺术家住所以及多种多样的功能空间，为Sinthian和周边地区的人们提供了发现新创造形式和培养技能的机会。作为一个集市场、教育、表演和会议于一体的场所，该中心是当地社区的枢纽，也是一个让住在这里的艺术家可以充分体验Sinthian社会的地方。

来自世界各地的画家、雕塑家、摄影家、作家、舞蹈编导、音乐家和舞蹈家被邀请到现场，并在Thread中心工作，但中心特别欢迎并鼓励当地和塞内加尔的艺术家参与。艺术家们把这个世界上很少有人访问的地方作为他们沉思冥想的地点，从而为西非这一地区带来更多的来自国际的关注与赞赏。有关这种文化交流的第一步发生在2013年，来自韦恩·麦格雷戈的兰登舞蹈团的舞者天才在Sinthian开办了一系列工作室。

办公室设在纽约的广受好评的建筑师森俊子无偿为这个项目设计了一座建筑，并赢得了AIA纽约分会奖，入选2014年威尼斯建筑双年展。该建筑利用当地材料进行建造，当地建筑商分享了他们使用竹子、砖和茅草工作的熟练知识。这些传统技术与森俊子的创新设计相结合。惯用的斜屋顶被反转，能够在降雨过程中收集雨水，可满足村民约40%的生活用水需求。

约瑟夫和安妮·阿尔伯斯基金会总监尼古拉斯·福克斯·韦伯，对Thread中心的精神做出评价，"约瑟夫和安妮·阿尔伯斯在用他们的名字成立基金会时就表明，他们的宗旨是'通过艺术来启示和召唤愿景'。他们把创造的行为和发现的乐趣作为战胜困难、保持平衡和充满希望的最佳手段。安妮·阿尔伯斯经常谈到生活的基础是'从零开始'，约瑟夫也经常赞美建筑实验所带来的奇妙体验"。Thread中心就是以与上述价值观相一致的方式建成的。地方领导Magueye Ba医生和美国非营利性组织AFLK已经携手合作运营Sinthian的医疗中心，建设Sinthian第一所幼儿园，资助教师工资，并帮助社区开展新的农业实践。现在，在约瑟夫和安妮·阿尔伯斯基金会的帮助下，村里建成了一个新的设施，补充了原有的基础设施，必将进一步促进该区域的创意发展。

Thread Artist Residency and Cultural Center

The rural village of Sinthian in south-eastern Senegal will be the setting for an exciting new cultural center, conceived and funded by the Josef and Anni Albers Foundation in Connecticut in collaboration with a local leader in Sinthian. Thread offers artist residences alongside a diverse range of programs

1. 蓄水池
2. 客房
3. 主管住宅
4. 门卫室/储藏室
5. 停车场
6. 太阳能板
7. 新道路
8. 原有道路
9. 医疗诊所
10. 住房

1. water reservoir
2. guest house
3. director residence
4. guard house/storage
5. parking
6. solar panels
7. new path
8. existing path
9. medical clinic
10. housing

that provides the people of Sinthian and the surrounding region with the opportunity to discover new forms of creativity and cultivate their skills. As a venue for markets, education, performances and meetings, the center is a hub for the local community and a place where the resident artists can have a truly meaningful experience of Sinthian society. Painters, sculptors, photographers, writers, choreographers, musicians and dancers from around the world are invited to live and work at Thread but the center particularly welcomes and encourages the participation of local and Senegalese artists. With this rarely-visited area of the world as their muse, these artists can inspire a greater international appreciation for this part of West Africa. The first step towards this spirit of cultural exchange was made in 2013, when gifted dancers

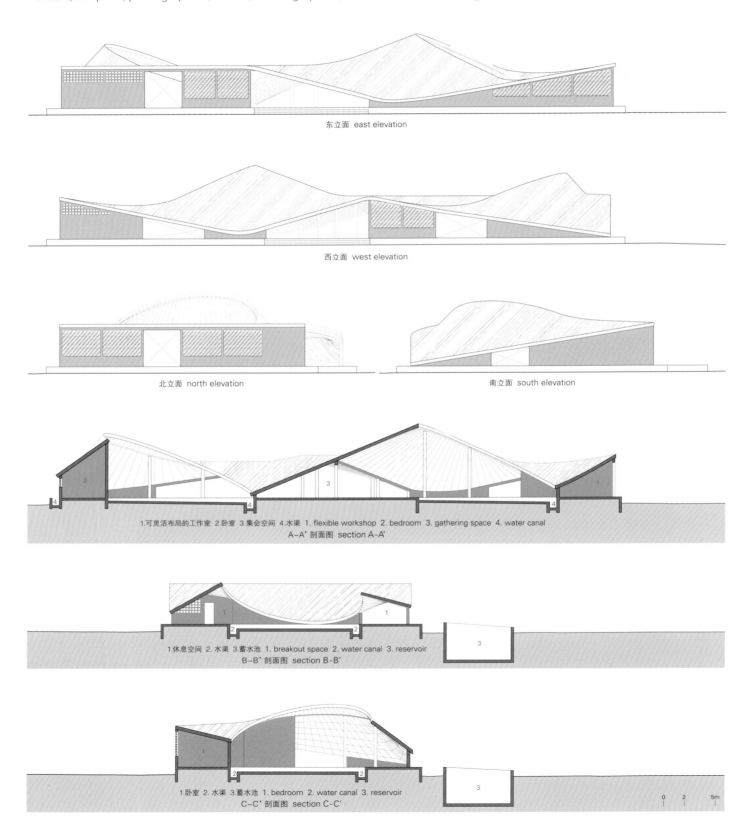

local materials and construction techniques serve as inspiration of roof structure

bamboo lashing techniques in Japanese traditional construction can improve overall construction and durability of design

traditional bamboo substructure

材料与结构
material and structure

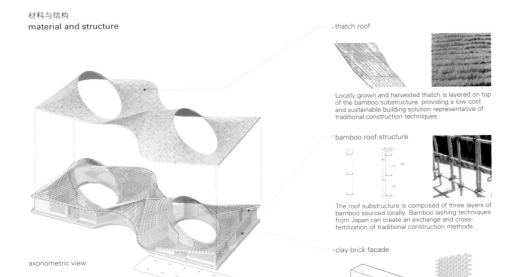

axonometric view

- thatch roof

Locally grown and harvested thatch is layered on top of the bamboo substructure, providing a low cost and sustainable building solution representative of traditional construction techniques.

- bamboo roof structure

The roof substructure is composed of three layers of bamboo sourced locally. Bamboo lashing techniques from Japan can create an exchange and cross-fertilization of traditional construction methods.

- clay brick facade

Clay bricks will be formed on-site by local villagers, enhancing the participation of the local community in the construction of the cultural center.

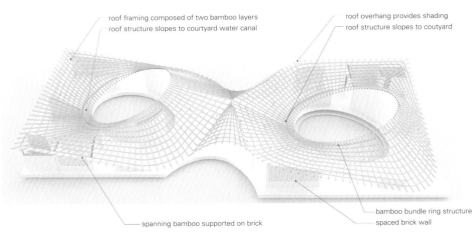

- roof framing composed of two bamboo layers
- roof structure slopes to courtyard water canal
- roof overhang provides shading
- roof structure slopes to coutyard
- spanning bamboo supported on brick
- bamboo bundle ring structure
- spaced brick wall

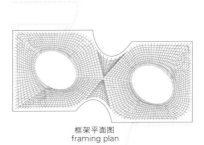

框架平面图
framing plan

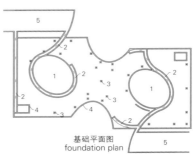

基础平面图
foundation plan

1. 倾斜庭院
2. 倾斜水渠
3. 柱子与柱脚
4. 连续基础
5. 蓄水池

1. sloped courtyard
2. sloped water canal
3. column and footing
4. continuous foundation
5. water reservoir

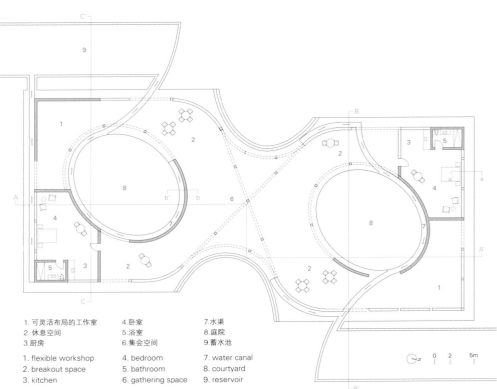

1. 可灵活布局的工作室
2. 休息空间
3. 厨房
4. 卧室
5. 浴室
6. 集会空间
7. 水渠
8. 庭院
9. 蓄水池

1. flexible workshop
2. breakout space
3. kitchen
4. bedroom
5. bathroom
6. gathering space
7. water canal
8. courtyard
9. reservoir

0 2 5m

from Wayne McGregor's Random Dance running a series of workshops in Sinthian.

Acclaimed New York-based architect Toshiko Mori has worked on this project pro-bono, designing a building that has already won an AIA New York Chapter award and was selected for the 2014 Venice Architecture Biennale. The building is constructed using local materials and local builders have shared their sophisticated knowledge of working with bamboo, brick, and thatch. These traditional techniques are combined with design innovations by Mori. The customary pitched roof is inverted and will be capable of collecting approximately 40% of the villagers' domestic water usage in fresh rainfall. Nicholas Fox Weber, director of the Josef and Anni Albers Foundation comments on the ethos of Thread, "When Josef and Anni Albers created the Foundation that bears their names, they stated its purpose to be 'the revelation and evocation of vision through art'. They regarded the act of creation and the pleasures of seeing as the greatest means to combat hardship and provide balance and hope. Anni Albers often spoke about 'starting at zero' as essential in life and Josef often extolled the wonders of experimentation." Thread has been built in accord with these values. Local leader, Dr. Magueye Ba and U.S. based non-profit organisation, American Friends of Le Korsa (AFLK) have worked together in running Sinthian's medical center, building its first kindergarten, funding its teachers' salaries and helping the community initiate new agricultural practices. Now, with the help of The Josef and Anni Albers Foundation, the village has a new facility that complements the existing infrastructure, and promises to stimulate further creativity in the region.

项目名称：Artist Residency and Cultural Center Thread
地点：Sinthian, Senegal
建筑师：Toshiko Mori Architect
合作者：Jordan MacTavish
主管：Nick Murphy
总经理：Moussa Diogoye Sene
支持：The Josef and Anni Albers Foundation, American Friends of Le Korsa (AFLK)
项目咨询：Koyo Kouoh
建筑面积：1,048m² / 有效楼层面积：1,048m²
设计时间：2014 / 施工时间：2014 / 竣工时间：2015
摄影师：
©AFLK and Thatcher Cook(courtesy of the architect)-p.58, p.59, p.61, p.63, p.69
©Iwan Baan(courtesy of the architect)-p.56~57, p.64~65, p.66~67

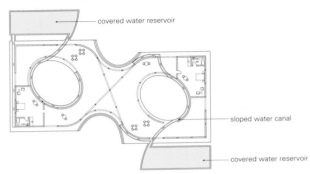

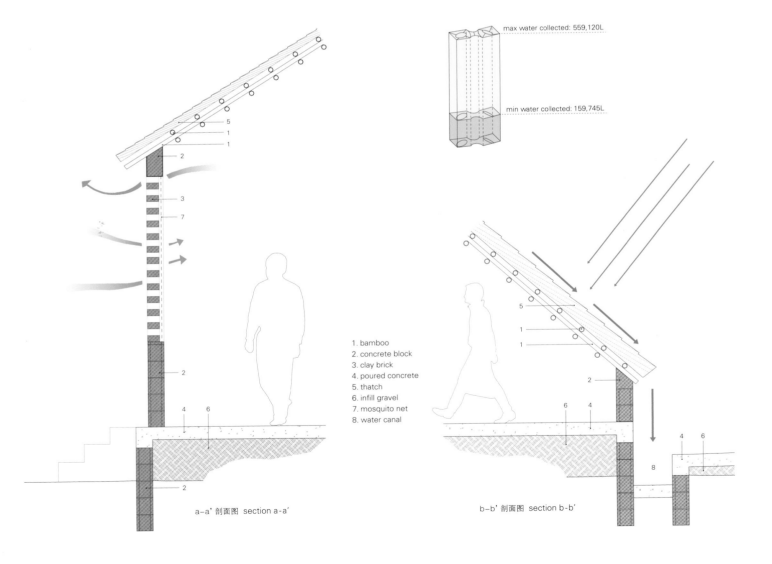

1. bamboo
2. concrete block
3. clay brick
4. poured concrete
5. thatch
6. infill gravel
7. mosquito net
8. water canal

a-a' 剖面图 section a-a'   b-b' 剖面图 section b-b'

牛背山志愿者之家
dEEP Architects

　　牛背山志愿者之家是一个公益设计项目，为年轻的志愿者们在大山里盖一座房子，地点在四川省泸定县蒲麦地村。它不仅为有需要的旅客提供帮助，还可以为村里的老人、儿童提供服务和帮助。为了保证这个社会公益项目的资金平衡，这座房子也需要具有青年旅社的功能。

　　改造前的房子是一个传统的民居，木结构坡屋顶上的瓦片已经破碎，门前有一个被当地人叫作"坝子"的平台。房子破旧不堪，被厚重的墙体分割成几个昏暗的小房间。屋顶阁楼也老旧破败，没有厨房，也没有浴室。

　　我们的改造策略是在完善基本使用功能的前提下，使结构更具有开放性与公共性，可为更多的人群服务，从建筑空间与结构上创新的同时，又使其与周围环境和传统文化相协调。因此我们保留并加固了内部的木结构，拆除了面向坝子的厚重墙体和内部隔墙，使一层完全开放，作为最重要的公共空间，人们可在这里阅读、会面、喝饮料等。钢网架玻璃门可以存储木柴，完全打开时，室内外将融为一体。

　　在这个项目中，我们使用最基本的建筑材料（比如与原来的建筑语言相同的石墙、坡屋顶和绿瓦）和最常见的建筑技术，并尽可能地使用当地的人力资源。同时，在这一过程中还采用了现代数字逻辑和设计策略。

　　当你面对主立面从左看向右时，会发现一个有机形态的屋顶与背

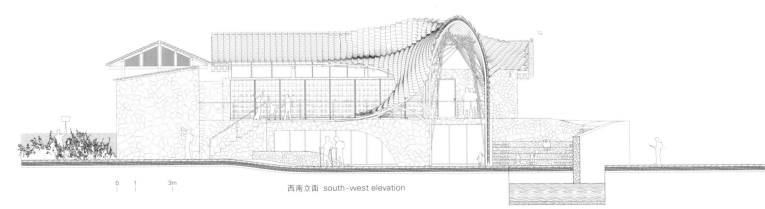

西南立面 south-west elevation

后的山和云融合在一起，发现由传统艺术完美转变到了现代艺术，甚至转变到了对未来的探索。

Cattle Back Mountain Volunteer House

The Cattle Back Mountain Volunteer House is a social welfare project dedicating to building a house for young volunteers in the mountain of Pumaidi Village, Luding County, Sichuan Province. It is not only to provide help for those travelers in need, but also the elderly and children in the village. The house will also serve as a Youth Hostel so as to keep the financial balance of the social project.

Upon renovation, it is an old traditional folk house, with wood pitched roof and broken tiles, facing with a platform(Bazi) in front, that is fragmented and shaded as dark rooms by thick walls. The loft on the rooftop is old and shabby, with no kitchen or bathroom inside.

1.民舍 2.废弃小学 3.村委会 4.农舍旅馆 5.废弃房屋
1. folk house 2. disused elementary school 3. village government 4. cottage hostel 5. disused house

项目名称：Cattle Back Mountain Volunteer House
地点：Sichuan, China
建筑师：dEEP Architects
总建筑师：LI Daode
功能：volunteer house, youth hostel, book and cafe bar, workshop spaces, medical room
用地面积：550m²
建筑面积：300m²
有效楼层面积：300m²
设计时间：2014.7 / 施工时间：2014.7~11
摄影师：courtesy of the architect

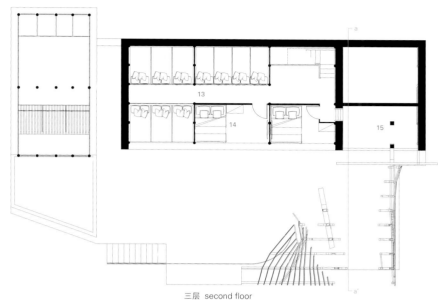

三层 second floor

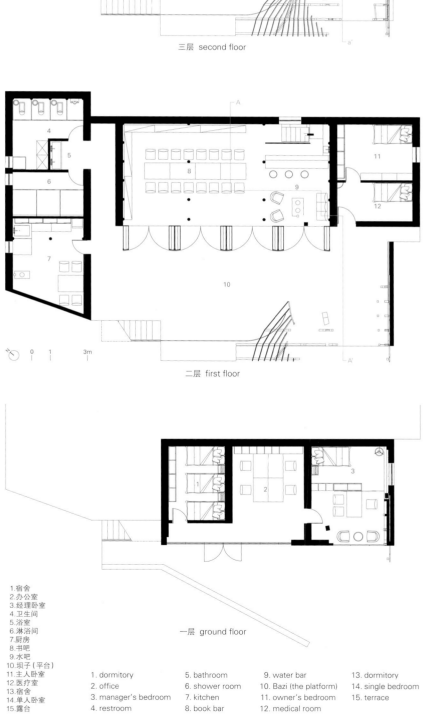

二层 first floor

一层 ground floor

1. 宿舍
2. 办公室
3. 经理卧室
4. 卫生间
5. 浴室
6. 淋浴间
7. 厨房
8. 书吧
9. 水吧
10. 坝子（平台）
11. 主人卧室
12. 医疗室
13. 宿舍
14. 单人卧室
15. 露台

1. dormitory
2. office
3. manager's bedroom
4. restroom
5. bathroom
6. shower room
7. kitchen
8. book bar
9. water bar
10. Bazi (the platform)
11. owner's bedroom
12. medical room
13. dormitory
14. single bedroom
15. terrace

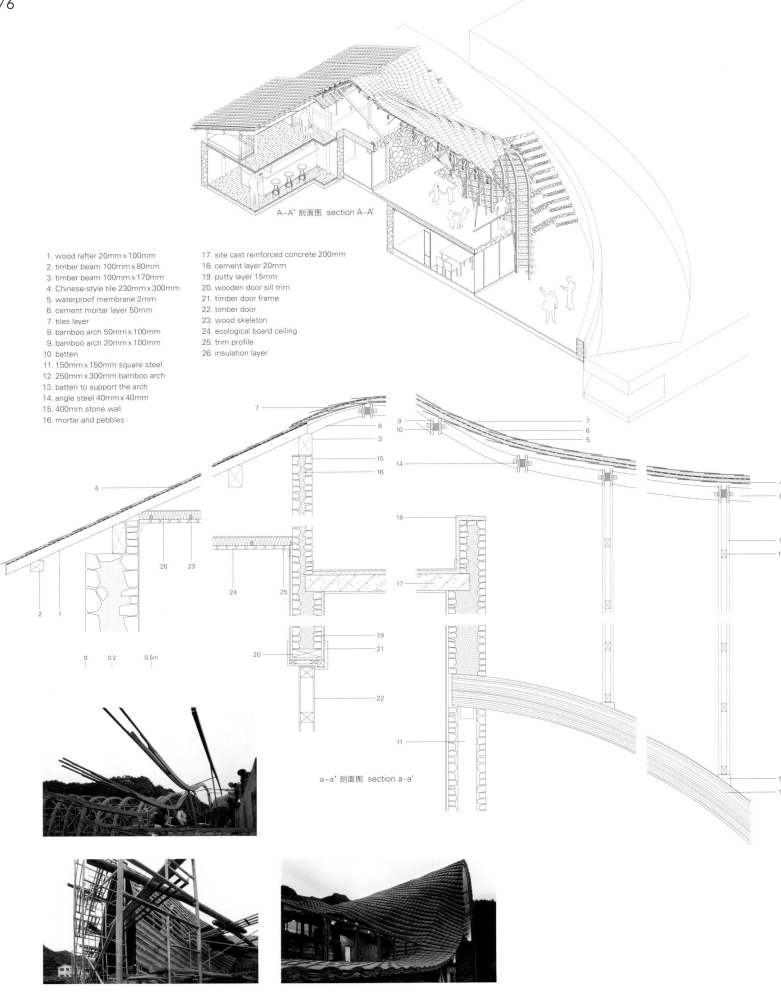

A-A' 剖面图 section A-A'

1. wood rafter 20mm×100mm
2. timber beam 100mm×80mm
3. timber beam 100mm×170mm
4. Chinese-style tile 230mm×300mm
5. waterproof membrane 2mm
6. cement mortar layer 50mm
7. tiles layer
8. bamboo arch 50mm×100mm
9. bamboo arch 20mm×100mm
10. batten
11. 150mm×150mm square steel
12. 250mm×300mm bamboo arch
13. batten to support the arch
14. angle steel 40mm×40mm
15. 400mm stone wall
16. mortar and pebbles
17. site cast reinforced concrete 200mm
18. cement layer 20mm
19. putty layer 15mm
20. wooden door sill trim
21. timber door frame
22. timber door
23. wood skeleton
24. ecological board ceiling
25. trim profile
26. insulation layer

a-a' 剖面图 section a-a'

Our design strategy is, while improving the basic programs and functions, to make the architecture more open and public, capable of serving more people, and blend the new building equipped with architectural and structural innovation with the surrounding environment and its traditional culture. So we kept and strengthened the internal wood structure, got rid of the thick walls facing with the Bazi and its partitions to open up the first floor and serve as an important public space where people can read, meet and grab a drink. The steel net framed glass wall can be used to store firewood, and when it opens completely, it merges the interior and exterior into one.

In the project, we used the most basic architecture materials (stone walls, pitched roofs and green tiles that consists with the original language), and the most common building technique, and we try our utmost to make full use of the local human resource. Meanwhile, the modern digital logic and design strategies are implemented in the process.

Facing the main facade, from left to right, you will be able to see an organic roof shape merging with the mountain and clouds at its back, and the perfect transition from traditional art to modern art, or even the exploration of the future.

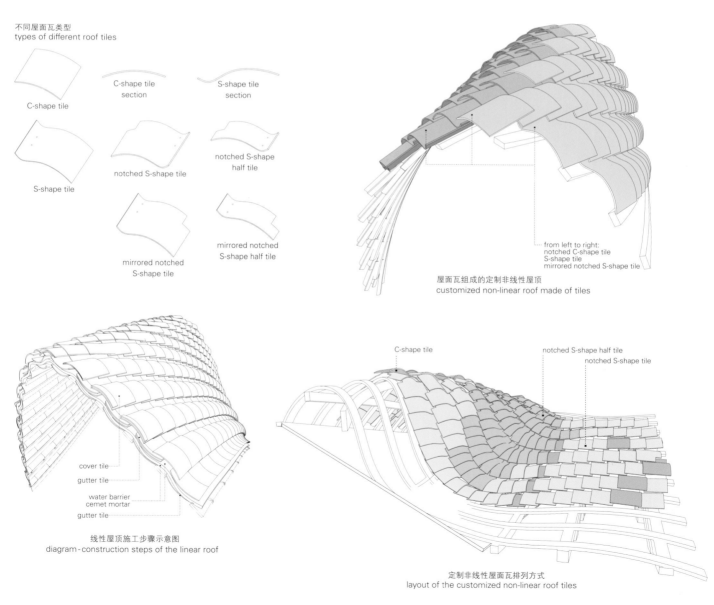

不同屋面瓦类型
types of different roof tiles

C-shape tile

C-shape tile section

S-shape tile section

S-shape tile

notched S-shape tile

notched S-shape half tile

mirrored notched S-shape tile

mirrored notched S-shape half tile

from left to right:
notched C-shape tile
S-shape tile
mirrored notched S-shape tile

屋面瓦组成的定制非线性屋顶
customized non-linear roof made of tiles

cover tile
gutter tile
water barrier
cemet mortar
gutter tile

线性屋顶施工步骤示意图
diagram - construction steps of the linear roof

C-shape tile
notched S-shape half tile
notched S-shape tile

定制非线性屋面瓦排列方式
layout of the customized non-linear roof tiles

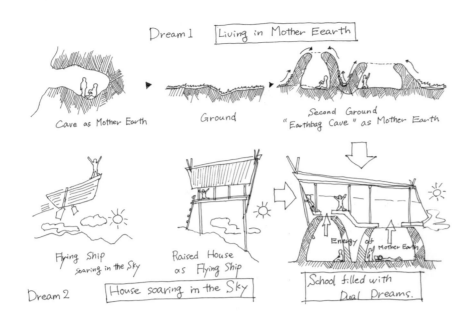

漂浮在天空中的孤儿学校
Kikuma Watanabe

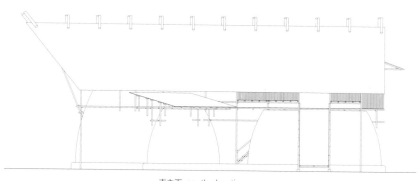

南立面 south elevation

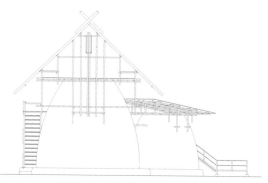

西立面 west elevation

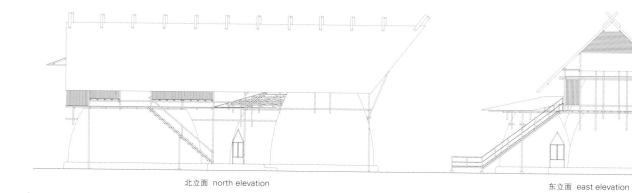

北立面 north elevation

东立面 east elevation

0 1 2m

设计这所学校的目的是帮助减轻泰国Shangkhaburi村的贫困状况，该村庄位于缅甸边境附近。这个区域有很多移民和孤儿，这些移民贫穷到没有能力抚养他们自己的孩子。

我们想要为这些孩子提供一个美好的未来，希望学校能实现他们的梦想。因此，最初老师让孩子们画出他们梦想中的学校建筑。其中一个学生画了一艘飞船。我们修改了他的想法，并把图画转变为建筑设计。图画被修改成两个主要组件：地面上的圆形土袋体量以及覆盖竹草屋顶的轻钢结构。

土袋体量的圆顶被设想成为船提供地球能量的"发射台"，上部钢结构建筑是在天空中翱翔的船。圆形的体量营造出温馨的内部环境，给身处祈祷穹顶和土地教室的孩子们带来了舒适感。功能区上面的"漂浮"层用作佛学教室和学习区。微风吹过茅草屋顶，给人一种身在一条船上的感觉。楼上通过两个洞口连接下方土袋的圆顶。

自建成以来，学校已成功地成为社区居民学习、玩耍和日常祈祷的场所。他们的梦想被诠释成建筑的形式，这有助于引导孩子们走向美好的未来。

Floating in the Sky School for Orphans

This school aims to help alleviate poverty in Shangkhaburi village, Thailand, located near the border of Myanmar. There are a lot of immigrants and orphans in this area. They are so poor that they cannot raise their children themselves.
Wanting to provide a good future for these kids, we hope that the school will realize their dreams. So, at first the teacher asked the children to draw the dream of the school building. One of them drew a flying ship. We adapted his idea, and translated the drawing into architectural design. The image was adapted into two main components: the round, earthbag

项目名称：School floating in the Sky – School for Orphans in Thailand
地点：Shangkhaburi village, Kanchanaburi pref, Thailand
建筑师：Kikuma Watanabe
设计公司/结构设计：D Environmental Design System Laboratory
施工合伙人：D Environmental Design System Laboratory,
Local carpenters (Garian who are minor mountain people), Children of this School (orphans from Myanmar), Students of Kochi University of Technology
客户：Kagayaku Inochi (glorious life), a Japanese NPO
用地面积：20,080.7m²
建筑面积：125.1m²
有效楼层面积：104.60m²
工程造价：USD15,000
委托时间：2012.2~4
材料：earthbag domes _ earth in the site, outside wall _ finishing with mud plaster, inside wall _ finishing with traditional cloth tied by bamboo frame, light steel building_ steel pipe used for scaffold, floor _ bamboo cut down around the site, roof _ grass roof manufactured in Shangkhaburi village
设计时间：2012.4~12
施工时间：2013.1~8
竣工时间：2013.8
摄影师：courtesy of the architect

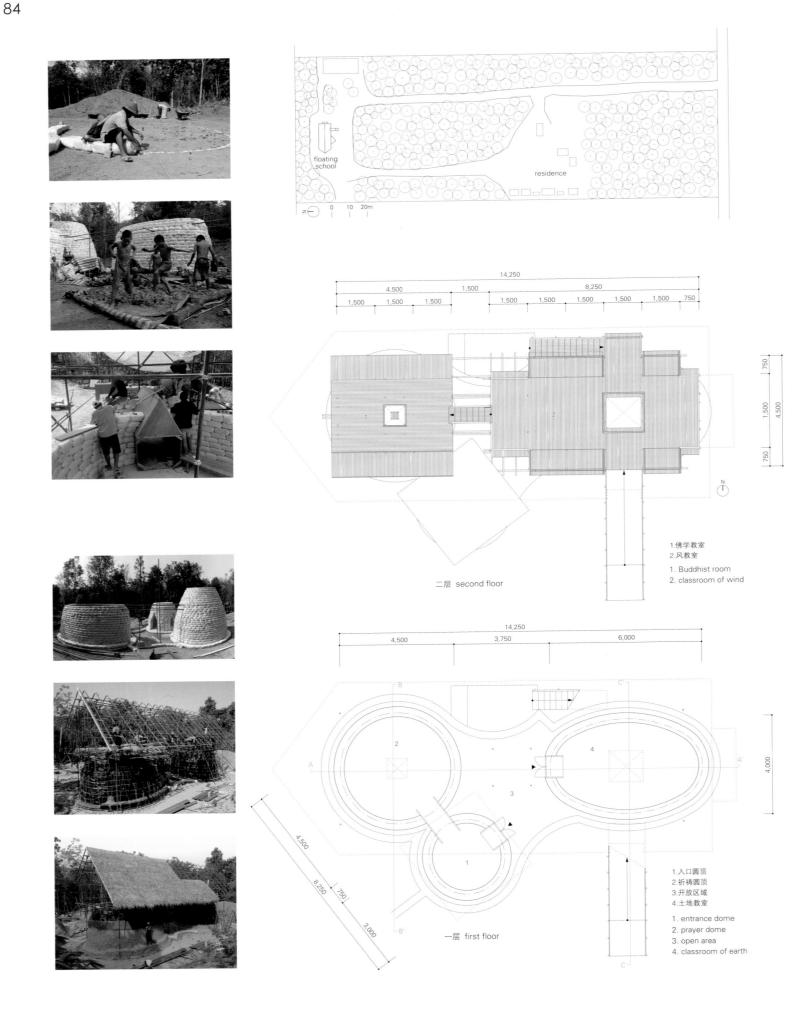

二层 second floor

1.佛学教室
2.风教室

1. Buddhist room
2. classroom of wind

一层 first floor

1.入口圆顶
2.祈祷圆顶
3.开放区域
4.土地教室

1. entrance dome
2. prayer dome
3. open area
4. classroom of earth

volumes on the ground and the other, a light steel structure finished with bamboo and a grass roof.

The earthbag domes were thought of as a "launching pad" that supplies the ship with the energy of Mother Earth and the upper steel building is the ship that is soaring in the sky. The round volumes create a warm interior, fostering a sense of comfort for the children in prayer dome and classroom of earth. The floating level above functions as a Buddhist room and learning area. A gentle breeze flows through the thatched roof, giving the feeling of being in a ship. The upper floor connects the lower earthbag domes through two openings.

Since its completion, the school has become a successful place for the community to enjoy studying, playing, and praying in their everyday. The interpretation of their dreams into architectural form establishes a foundation to help lead the children to a brighter future. Kikuma Watanabe

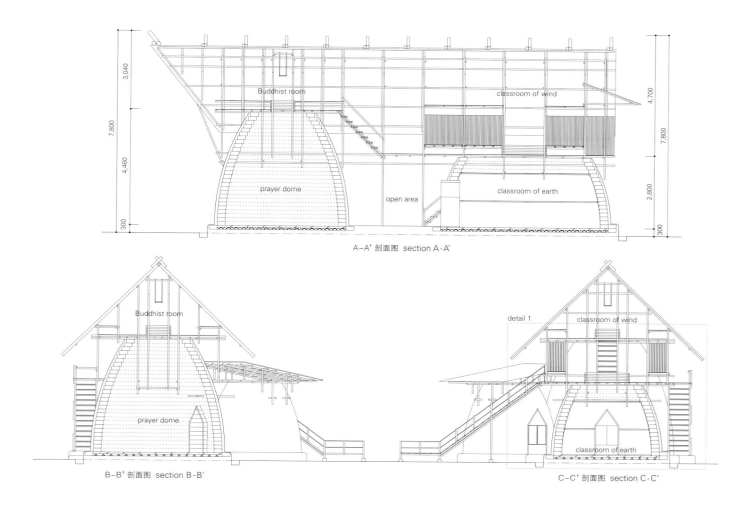

A-A' 剖面图 section A-A'

B-B' 剖面图 section B-B'

C-C' 剖面图 section C-C'

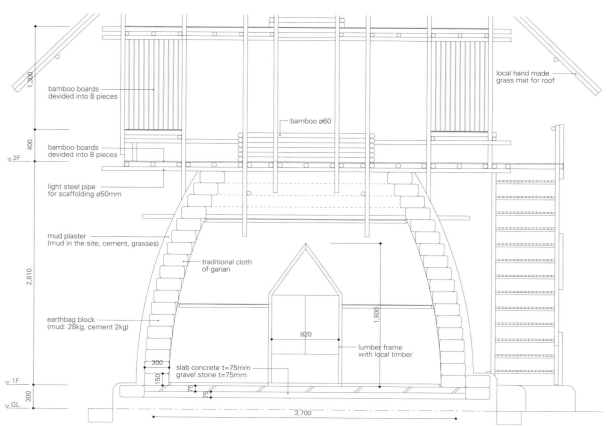

详图1 detail 1

Trika别墅是一座占地425m²的奢华住宅，拥有5间卧室和5间浴室，是泥土和竹子建造的当代建筑的典范。

巨大的竹屋顶像一把伞笼罩着整座房屋，阻挡着阳光和雨水，但是并没有封闭住房间，而是创建了一个与外部相连的开放的自然通风空间，因而不需要空调。每个房间都有一个蚊帐似的帐篷遮盖，取代了天花板，浴室采用半开放式设计——水槽和卫生间在屋顶下，淋浴则在开放的天空下。

三间卧室、一间露天的起居室和一间主卧室，三者各占一边（起居室位于短边），呈U形围绕着中央泳池。起居室／餐厅与宽敞的厨房相邻，与厨房紧邻的是保姆房与房屋后院。

所有的墙采用Adobe砖料建造，地面则是由夯土或黏土与混凝土混合而成。因此与传统的钢筋结构或混凝土结构住宅相比，这种房屋的碳排放量减少了大约90%，同时也没有放弃现代奢华的生活方式。

在Trika别墅中，因为室内外一直相连，所以自然与生活浑然一体。

Trika Villa

Trika Villa, a 425m² luxury residence with 5-bedrooms and 5 bathrooms is a prime example of modern earth and bamboo architecture.

A big bamboo roof is hovering above the whole mansion like an umbrella, protecting it against sunshine and rain, but not sealing off the rooms, creating open, naturally ventilated spaces that are still connected to the outside and don't need air-conditioning. Instead of the ceiling, each room has a mosquito net tent cover and bathrooms that are half open – the

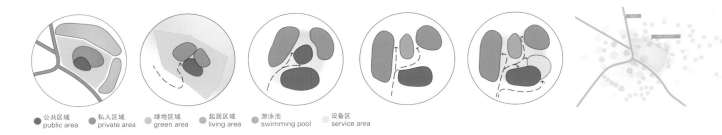

公共区域 public area ● 私人区域 private area ● 绿地区域 green area ● 起居区域 living area ● 游泳池 swimming pool ● 设备区 service area

Trika别墅

Chiangmai Life Construction

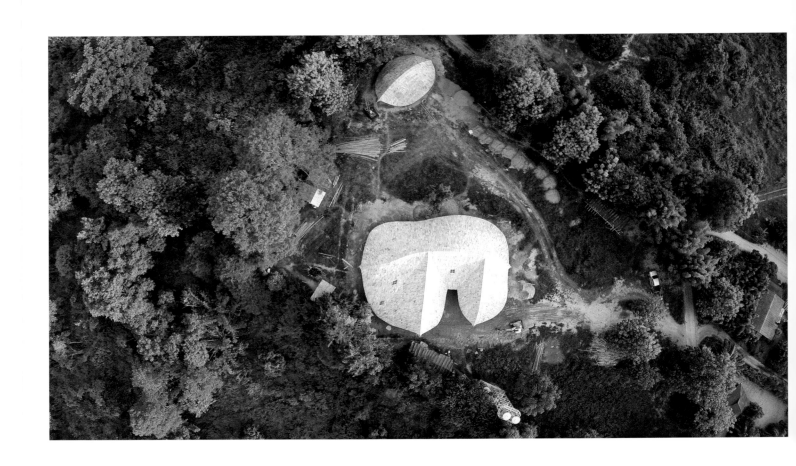

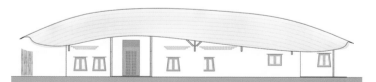
南立面 south elevation

西立面 west elevation

北立面 north elevation

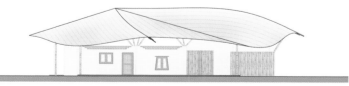
东立面 east elevation

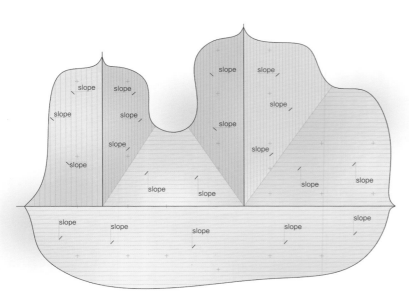

屋顶 roof

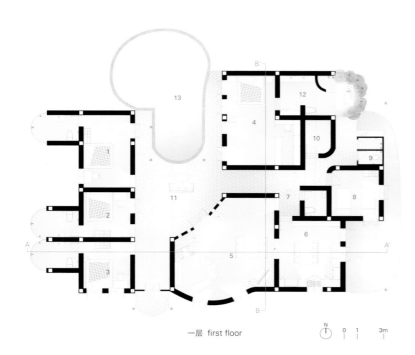

一层 first floor

1.卧室1&浴室 2.卧室2&浴室 3.卧室3&浴室 4.主卧 5.起居室 6.厨房 7.公共卫生间 8.保姆房 9.保姆卫生间 10.储藏室 11.起居区 12主人浴室 13.游泳池
1. bedroom 1&bathroom 2. bedroom 2&bathroom 3. bedroom 3&bathroom
4. master bedroom 5. living room 6. kitchen 7. public toilet 8. maid room
9. toilet for maid 10. store 11. living area 12. master bathroom 13. swimming pool

sink and toilet are under the roof, and the shower is under the open sky.

A central swimming pool is surrounded in a U shape by 3 bedrooms on one side, an open air living room on the small side and the master bedroom on the other side. A living/dining room is adjacent to the big kitchen, attached to which are the maid quarters and back of the house areas.

All walls are made of Adobe bricks and the floors of rammed earth or a clay-concrete mix. Thus the carbon footprint of this mansion compared to a conventional steel/concrete version was reduced by about 90 percent without having to forego a modern and luxurious lifestyle.

At Trika Villa nature automatically becomes part of life as inside and outside are always connected.

项目名称：Trika Villa
地点：Chiang Mai, Thailand
建筑师：Markus Roselieb, Tosapon Sittiwong _ Chiangmai Life Construction (CLC)
工程师：Phuong Nguyen, CLC
Project management / Construction: Chiangmai Life Construction (CLC)
占地面积：425m²
竣工时间：2014
摄影师：©Alberto Cosi(courtesy of the architect)

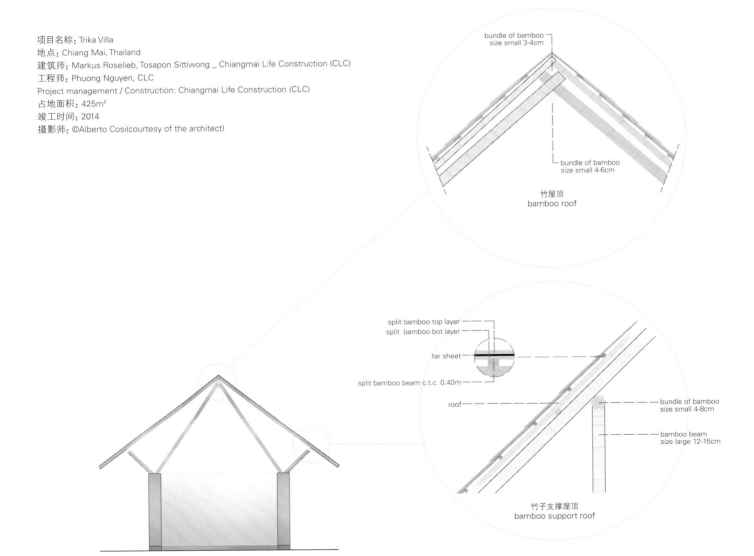

1. 卧室3&浴室 2. 起居室 3. 厨房 1. bedroom 3 & bathroom 2. living room 3. kitchen
A-A' 剖面图 section A-A'

1. 起居室 2. 主卧 1. living room 2. master bedroom
B-B' 剖面图 section B-B'

地方性与全球多样性 Regionalism and Global Diversity

紧凑的喀斯特房屋
Dekleva Gregorič Arhitekti

喀斯特地区曾一度被橡树所覆盖。威尼斯人大量地使用这些橡树在水边建立城市，他们让风吹去土壤，露出石灰岩地面。在这片风景中，建造狭小、紧凑甚至没有窗户的石头房屋成为一种流传至今的传统。

根据这样的传统，建筑师设计了这座紧凑的小型石头房屋，同时也满足了这个年轻家庭和现代技术准则的要求。建筑师对传统喀斯特石屋进行了重新定义，创造了房屋原型理念，这个原型就是紧凑的坡屋顶石头体量，以此来匹配这一地区的当代乡村生活。房屋被设计成一个整体的体量，插入的两个木质体量由中间的楼梯平台连接。

房屋一层主要当作公共的或半公共的空间，可以看到广阔的风景，而上层则非常私密，只能看到天空。房屋整个空间由插入其中的两个木质体量分隔开，一层的木质体量包括一间设有餐厅的厨房和浴室，主卧和儿童房则位于上层的木质体量中。房中房的设计构想是每一个卧室都是一个木质坡屋顶房屋，睡在其中，人们完全感觉是在自己的房子里，而不是在一个房间里。连接两个房屋的桥梁可作为游戏室使用。房屋设有三扇巨大的方形窗户，向西能看到外面山顶上的意大利教堂，向南则是一片森林，而东面是入口平台。

建筑师大量运用独创性技术，通过纹理、颜色、材质和陡峭的斜坡做出当代的、具体的诠释，对传统的喀斯特石屋顶进行了重新定义。立面和屋顶间的连接完全不可分割，它是传统喀斯特村庄形象的主要标志。

在建造过程中，把石头砌筑在混凝土中，墙体就能呈现出牢固的砌体结构外观。Aljoša的先驱建筑师于20世纪70年代晚期在喀斯特地区就运用过这种"低技"技术。立面层厚15cm，建造者首先铺设一层石块，让石块较为平整的一面对着框架，然后从这排石块后面浇筑混凝土。一到两天后移开框架，建造者最后去除多余的混凝土，露出石块表面。天花板是混凝土板屋顶结构——模板由厚木板制成。室内所有干燥的墙壁都由三层云杉木胶合板制成，板材涂抹了透明的油。水平板材是云杉木交叉层压木板。

房屋的设计凸显了现代与传统之间的关系，同时又开启了有关这种无特征的传统建筑具有怎样的特色这样的话题。这种设计既源于这样的传统建筑，同时又建立了现代诠释与传统受限定的综合体之间的关系。

Compact Karst House

The region of the Karst was once covered with Oak trees that Venetians extensively used to build up the City on water. They left the wind to peel off the earth revealing limestone ground. In this landscape, the tradition of small, compact, stony and almost windowless houses developed and has remained until today.

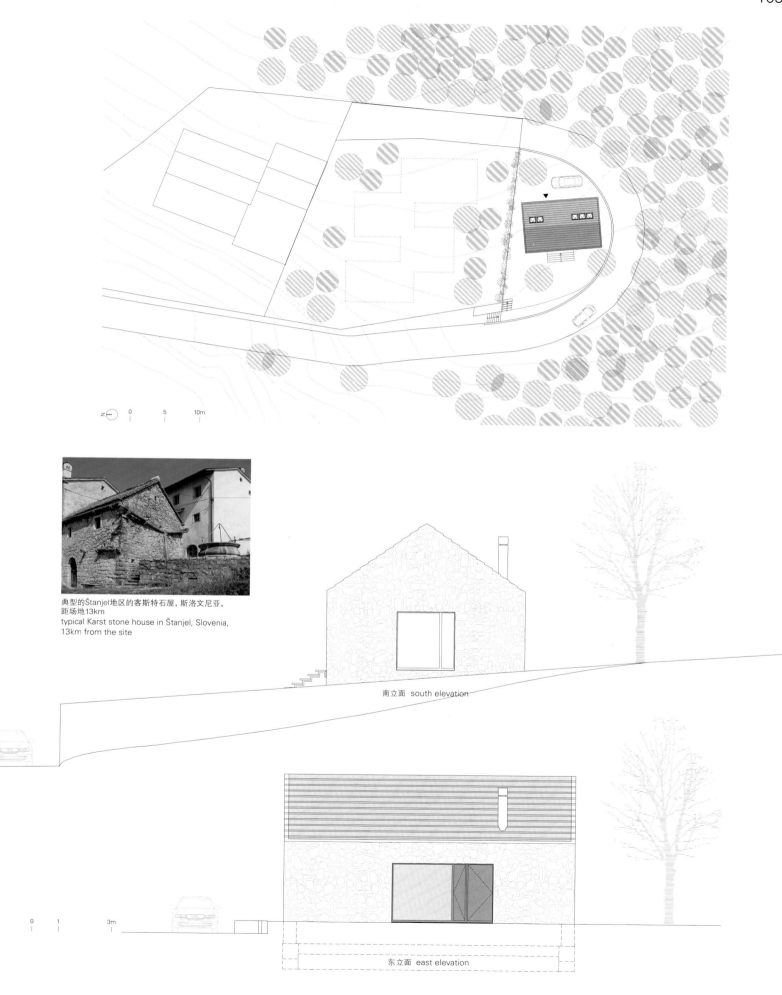

典型的Štanjel地区的客斯特石屋，斯洛文尼亚，距场地13km
typical Karst stone house in Štanjel, Slovenia, 13km from the site

南立面 south elevation

东立面 east elevation

Following this tradition determined the design of a small compact stony house, corresponding to the needs of the young family and current technological principles. The redefinition of a traditional stony Karst house led to the concept of the proto-house as a compact, stony, pitched roof volume for contemporary countryside living in this region. The house is conceived as a monolithic volume with two inserted wooden volumes connected with an interim landing.

The ground floor operates mostly as a public or semi-public space with multiple grand landscape views, whereas on the other hand the upper floor stands very private, with sky views only. The space is divided with two inserted wooden volumes which contain a kitchen with dining and bathroom on the ground floor, and a master bedroom and children's room on the upper floor. The house in a house concept allows each bedroom to perform as a primarily wooden pitched house, where one literally feels like sleeping in his own house and not a room. The bridge connecting both houses acts as

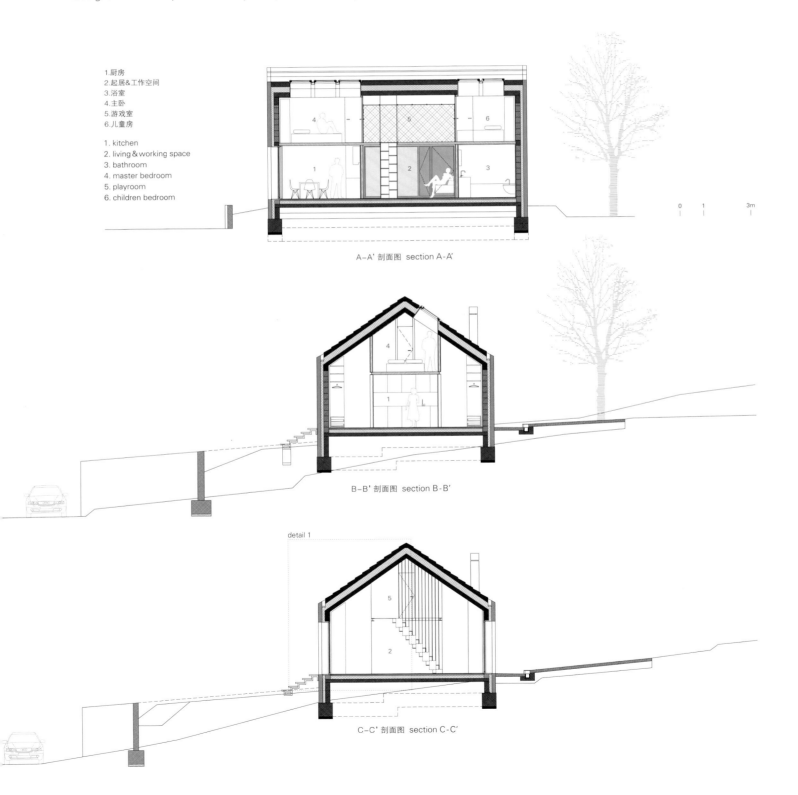

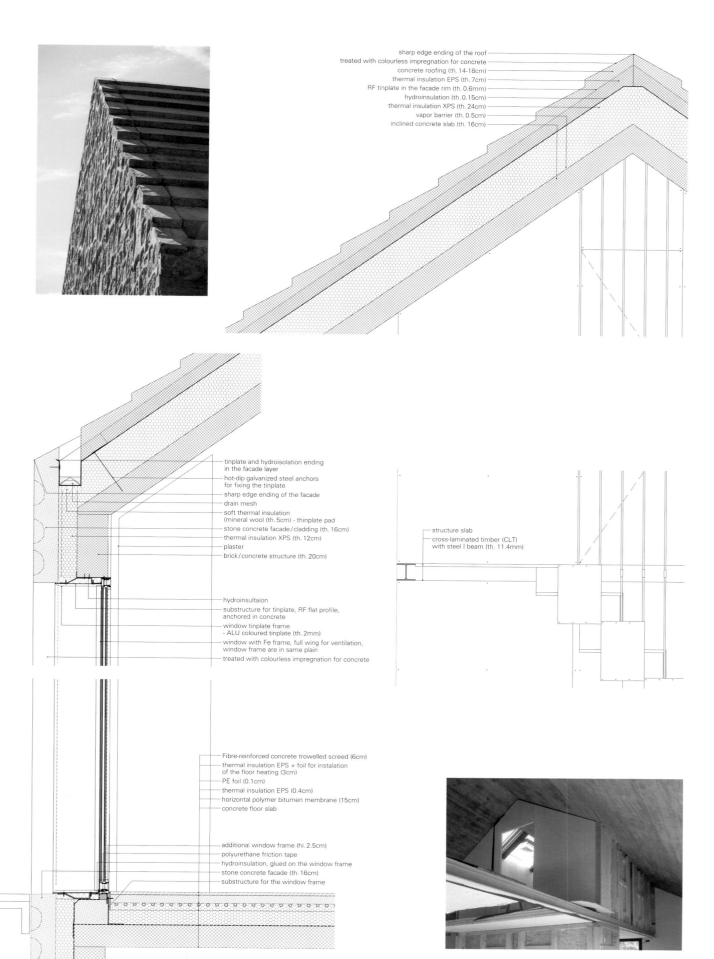

详图1 detail 1

playroom. The house has three large square windows which look out on views towards a hilltop church in Italy to the west, forest to the south and entrance platform to the east.

The redefinition of a traditional stony Karst roof, with its texture, colour, material and its steep inclination is executed as a contemporary concrete interpretation, with intense technological ingenuity. A materially inseparable connection between the facade and the roof is a key allusion to the image of the traditional Karst village.

The stone, set into the concrete during the casting process, gives the walls the appearance of having a solid masonry construction – this "low tech" technique previously employed by Aljoša's architect in the late 1970s in the Karst region. It is a 15 centimetre thick facade layer where the builder first lays down a row of stones, with the more or less flat side of the stone against the framework, and then pours concrete behind the row of stones. When the framework is removed, after one or two days, the builder would eventually remove excessive amounts of concrete to open the surface of the stone. The ceiling is the concrete slab of the roof construction – the formwork was done out of wooden planks. All interior dry walls are made of three layer spruce plywood panels, oiled with transparent oil. The horizontal slab is spruce Cross Laminated Timber.

The design of the house addresses the relationship between the contemporary and tradition, and it opens up the question about the characteristics of the anonymous traditionally built architecture from which it originates and simultaneously establishes the relationship between contemporary interpretation and the traditionally conditional domain of synthesis.

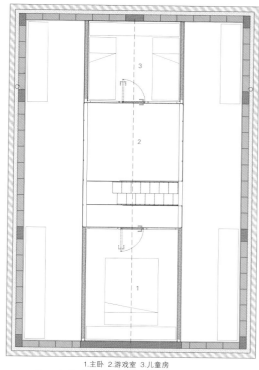

1.主卧 2.游戏室 3.儿童房
1. master bedroom 2. playroom 3. children bedroom
二层 second floor

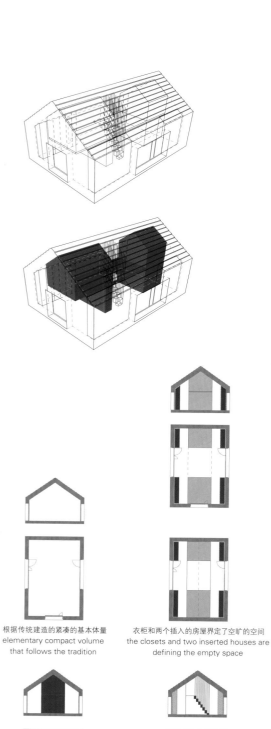

根据传统建造的紧凑的基本体量
elementary compact volume
that follows the tradition

衣柜和两个插入的房屋界定了空旷的空间
the closets and two inserted houses are
defining the empty space

两个插入的体量由中间的楼梯平台相连
two inserted volumes
with interim landing

具有不同外观的多功能楼梯
the staircase is a multifunctional
object with different appearances

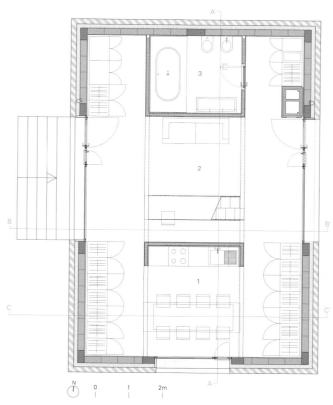

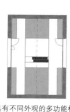

1.厨房 2.起居&工作空间 3.浴室
1. kitchen 2. living & working space 3. bathroom
一层 first floor

项目名称：Compact Karst House / 地点：Vrhovlje, Karst, Slovenia
建筑师：Dekleva Gregorič Architects
项目团队：Aljoša Dekleva, Tina Gregorič, Lea Kovič, Vid Zabel
客户：Borut Pertot, Pertot Showroom, S.L.R.
用地面积：330m² / 建筑面积：82m² / 有效楼层面积：92m²
设计时间：2011—2012 / 竣工时间：2014
摄影师：©Janez Marolt (courtesy of the architect)

P住宅
Budi Pradono Architects

　　P住宅是生活在附近的社区居民利用旧房屋用过的材料，最大化使用当地可以获得的材料，如周边地区的竹子、黏土、石块和砖，而建立的一座设有图书室的房屋，供当地人分享知识。由于业主的所有儿女都生活在其他国家或城市，因此他们需要建一个图书室用来集会。后来业主去世了，这座房屋就成为这个地区年轻创新者的创意分享空间，将他们的事业带上下一个台阶。

　　P住宅位于默巴布火山山脊，海拔2000m，周围围绕着其他一些山脉，如默拉皮火山和特落默约火山。该地区天气相当寒冷，平均气温在17~22℃。项目试图用多种方式来诠释爪哇村庄的房屋，某些空间里加入了高山形式，作为对周围山脉的诠释，同时也作为天窗使用，尽可能多地给建筑带来自然光。总体来说，竹子作为主体结构的材料，在项目周围极易获得。

　　项目在概念上试图通过开放和共享空间为家人带来童年的回忆。主浴室是社交空间，与其他区域互相影响。其他空间由流畅的空间相连，如厨房、休息室、食品储藏室、餐厅和家庭娱乐室，这些空间都是开放的，只有睡眠空间可独立控制，是封闭的，所有的公共空间也都是

开放的,朝向花园及其前面的热带森林。图书室或书房为椭圆形几何结构,采用亭子的形式单独建立。周围的社区居民可以阅读图书室里的图书。

项目的第二个主题是简约,因此建筑师通过其他方式使用了和周围地区房屋所用相同的材料,如砖、竹子和石块。从远处看去,这些建筑就跟周围村庄里的房屋一样。每个房间的门都循环利用了旧房屋的门,符合使用可回收材料的策略。建筑面向花园的一面呈现了最大限度的透明度。

场地内几乎所有现存的高大树木都被保留下来,用以突出这个地点的特色。花园中央种了一棵"啜泣"树,这棵能"治愈"各种疾病,因此这个地方对于周围社区变得更实用。房屋高度根据原有的轮廓进行调整。

这个项目展示了村庄房屋类型学。竹子结构在这里很常见,但使用竹子作为屋顶材料在该地区属于新的应用。

P House

P House is a house with a library for sharing knowledge to the local people, using used materials from the old house, maximizing the use of locally available materials (bamboo, clay, stone, and bricks) from surrounding area and built by the community who lives around the site. Since all the owner's sons and daughters live in other countries and cities, it needs a library for the people to gather around the house. Later after the owner passed away, it became a creative sharing space for young creative in the region to bring them to the next stage of their career.

P House is located at an altitude of 2,000m above the sea level and located on the ridge of Mount Merbabu and surrounded by several other mountains such as Mount Merapi and Mount Telomoyo. This area is quite cold with an average temperature of 17~22°C. This project seeks to interpret the Javanese village houses in multiplication and adds the form of mountains in some spaces as the interpretation of the surrounding mountains, which also serve as skylight to bring natural light as much as possible into the building. In general, bamboo as the material of main structure is easily found around the project.

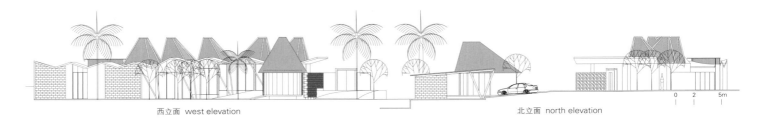

西立面 west elevation　　　　北立面 north elevation

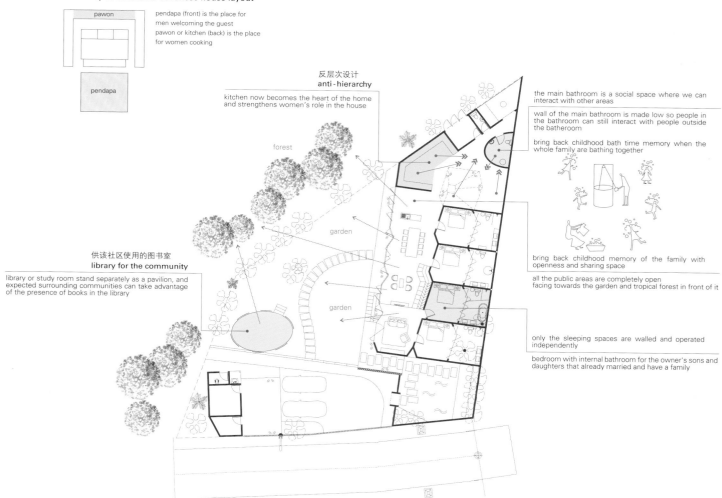

Conceptually this project is trying to bring a childhood memory of the family with openness and sharing space. The main bathroom is a social space which can still interact with other areas. Other spaces are connected by liquid space, which are kitchen, lounge, pantry, dining room and family room. All are completely open. Only the sleeping spaces are operated independently and remain closed. All the public areas are completely open, facing towards the garden and tropical forest in front of it. Library or study room is made with other geometry – an oval – and stands separately as a pavilion. Expected surrounding communities can take advantage of the presence of books in the library.

Simplicity is the second theme in this project so it uses the same materials as those used in the surrounding areas such as brick, bamboo and stone in another way. From a distance these buildings looked like houses in the surrounding villages. The doors used in each room are recycled doors from the old house, with a strategy of utilizing recycled materials. One side of the building is made as transparent as possible towards the garden.

Almost all existing large trees on the site are maintained as an attempt to emphasize a place. On the center of the garden, a "pule" tree is planted as this tree can be used to "heal" various diseases, so this place can be more useful to the surrounding community. Floor height is adjusted according to the existing contours.

This project is showing a typology of village house interpretation. Bamboo structure is dominant and significant, but the use of bamboo as a roofing material is something new there.

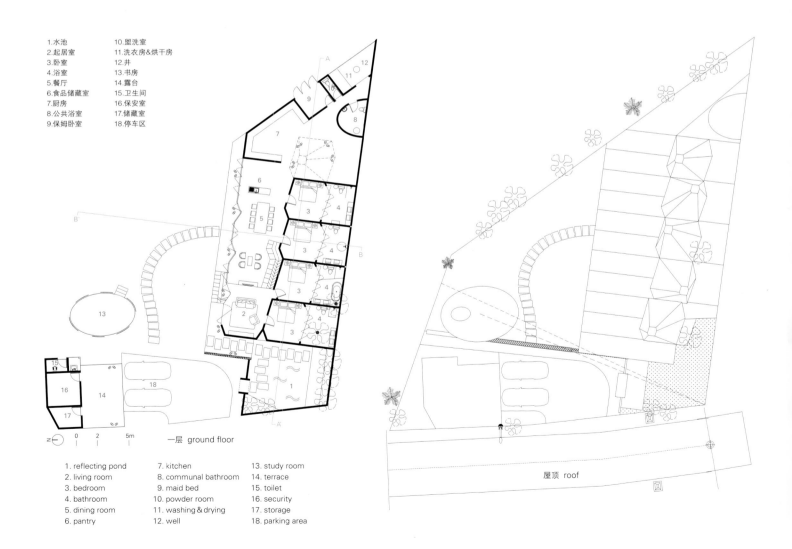

一层 ground floor

1. reflecting pond
2. living room
3. bedroom
4. bathroom
5. dining room
6. pantry
7. kitchen
8. communal bathroom
9. maid bed
10. powder room
11. washing & drying
12. well
13. study room
14. terrace
15. toilet
16. security
17. storage
18. parking area

屋顶 roof

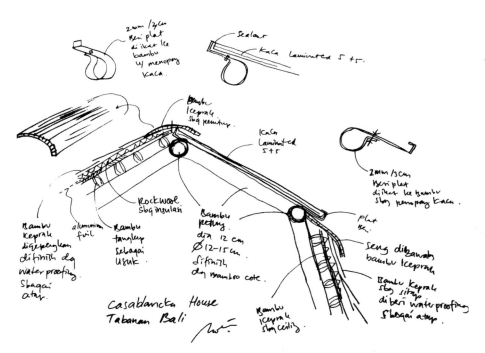

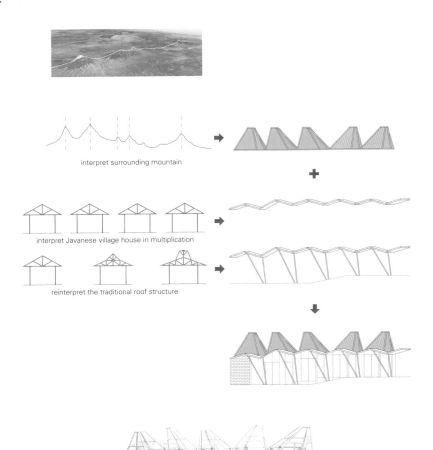

项目名称：P Hous
地点：Salatiga, Central Javwa, Indonesia
建筑师：Budi Pradono Architects
首席项目建筑师：Budi Pradono
参与设计建筑师：Stephanie Monieca, Arief Mubaraq
其他协助支持建筑师：Damicia Tangyong, Monica Selvinia, Indrawan Suwanto
模型制作：Daryanto
室内设计师：Budi Pradono Architects
土木工程承包商：surrounding community
客户：Wahyo Pradono and Eliana Rosalina Pradono
用地面积：714.86m² / 有效楼层面积：320.44m²
设计时间：2011 / 施工时间：2012 / 竣工时间：2014
摄影师：©Fernando Gomulya(courtesy of the architect)

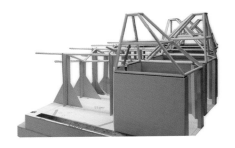

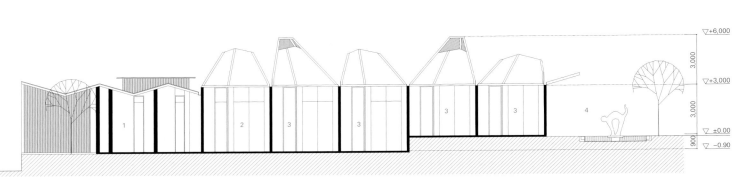

1.保姆卧室 2.厨房 3.卧室 4.水池　1. maid bed　2. kitchen　3. bedroom　4. reflecting pond
A-A' 剖面图　section A-A'

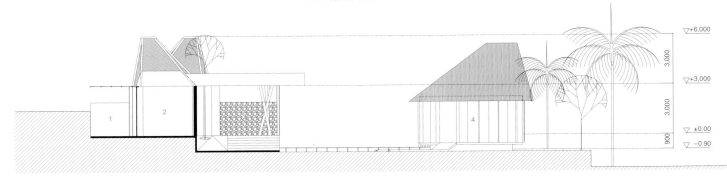

1.浴室 2.卧室 3.起居室 4.书房　1. bathroom　2. bedroom　3. living room　4. study room
B-B' 剖面图　section B-B'

居住在记忆里

游走在复兴与再利用之间

Dwelling in Me
between rehabilitation and reuse

  利用多余的空间一直就是居住的首选之一。地方建筑教会我们即使最简单的建筑如果不是完全依赖于原来存在的自然或者人工的建筑，也要与它们共存。在许多案例中，建筑师将有不同背景和功能需求的建筑结构比邻而建，都创造出非常有趣的混合体，这个混合体既没有必要将新旧区分开来，也不是既定建筑形式的连续演变。这些建筑很可能只是出于简单的需求而建，比如，为了节省材料而依附已存的墙体而建。但是在一个建筑潜能过剩的年代里，有些建筑来源于建筑师对美学品质的欣赏，例如，建筑所带有的记忆、时光、叙事或幻想性。

  一旦我们把家当成一个地方，使它超越简单的功能性而成为一个书写个人故事的地方，我们的住所就会成为一个混合体，承载过去和尚未书写的未来。这些建筑所体现出来的历史基本是个重写本，是不断进化永无止境的过程，在历史内讲述历史，依据历史讲述历史。与单独的和普遍的创造行为相反，我们所展现的房子仿佛从满是油污的画布中出现。这些房子所讲述的故事带有斑驳的铜锈。在这样的背景下，革新性成为适应性的一个问题；历史没有在盲目地对真实的学术追捧中消失殆尽，并且"当代的居住"成为拥抱历史魅力的展现体。

  因此，我们的居民游走在历史与现代之间。我们仿佛进入了一个反常的现实当中，行进过程的一瞥都能寻出旧日时光的精华，时间在科技、技术和想象之间保持了静止。显然这种历史介入现实的建筑方法需要遵循一定的要求、规则和涉及立法到学术要求的成规。但是很奇怪，我们还是感觉到这些建筑转变过程的最初规则仿佛在自身的条件内早已存在，好像它们自己孕育了自己的创新。从建筑学来讲，它需要光荣而谦逊的创造行为；从居住角度来讲，我们建造的不仅是一个生态学体系，更是一个新的居住习惯，一种新的生活方式。

Occupying redundant spaces has long been an option for dwelling. Vernacular architecture teaches us that even the simplest shelters co-exist with if not feed upon– pre-existing structures either natural or man-made. In many cases the juxtaposition between structures of different background or functional requirements can create an interesting mix, not necessarily an articulated division between old and new, but not nearly a consistent evolution of a given form. These structures may emerge out of simple necessity, such as building upon a pre-existing wall in order to save materials. But in an age with a superfluous potential for building, it seems they stem out of the simple aesthetic appreciation for qualities such as memory, time, narrative or fantasy.

Should we take "home" as a place beyond mere functionality to somewhere personal stories are written, our dwellings become a hybrid between the past and a yet-to-be-written future. They are, more or less, a palimpsest; histories told within histories, told upon histories, in a never ending process of constant evolution. Contrary to a singular and universal act of creation, the houses we present emerge out of an already soiled canvas, and the stories they tell are infested with patina. In this context, novelty becomes a question of adaptation: history is not exhausted in a blind academic adherence to authenticity, and "contemporary living" becomes a statement that embraces the allure of the past.

Thus our habitat becomes living in-between. It almost feels as if we enter a warped reality where glimpses of our progress seek out the essences of the past, and time stands suspended between technique, technology and the imagination. Obviously this type of intervention is required to follow a certain set of requirements, rules and preoccupations ranging from legislation to academic inquiry. Yet, it strangely feels as if the primary rule of these buildings' transformative process pre-exists within their own condition, as if they incubate the terms of their own regeneration. To name it architecture, it requires a gloriously modest act of creation; to inhabit it, we establish – more than a new ecology – a new habitus, a new way of living.

奥洛特某联排住宅_Row House in Olot/RCR Arquitectes
黛波拉某独栋住宅_Single Family House in Tebra/Irisarri-piñera S.L.P.
Scaiano旧石屋改造项目_Old Stone House Conversion in Scaiano
/Wespi de Meuron Romeo Architects
花园住宅_Garden House/Hertl. Architekten ZT GmbH

居住在记忆里：游走在复兴与再利用之间
Dwelling in Memory: between rehabilitation and reuse
/Angelos Psilopoulos

　　全世界的人口都在繁衍并不断地迁徙。这样的现象可以在有形和无形的人类建造行为中找到证据。部分建筑是住户在自己的土地上建造设计出来的。世代以来，这些房屋经历不同的使用、修缮，甚至受到自然力量、冲突、战争以及大灾难的破坏。然而重要的部分经受住了时间的洗礼，成为历史的见证，有的以废墟的形式呈现，有的保存完好，有的继续使用。

　　我们对待文化遗产的态度仍旧有待考证，主要的原则就是没有文化能够在失去历史的基础上探求未来。一旦我们意识到工业革命带来的科技进步给了我们力量，让我们以前所未有的速度来利用文化和自然环境，所有的一切都会变得有意义；面对那些不可抗力时我们勇敢面对，不会像以前那样被动承受（自然灾害、疾病、人类因为自身局限而对居住地选择的局限）。简言之，我们现在在技术上有能力决定我们是遵从自然所给与的，还是迅速打破现状。就居住而言，我们可以建我们喜欢的任何建筑，在任何地方建，根据我们的意愿、需求和希望来改造景观，使用任何材料和技术，而不用受特定地方的限制。我们可以随着我们的意愿而居住。

　　问题随之就变成了"为什么"而不是"怎么样"。为什么我们这样或那样建筑房屋？为什么我们选择世界上这块土地，或那块土地？为什么我们选择这种或那种风格来建筑？当人们享受某种程度的自由后，总会面对空虚盲目。为了表达的目的，美学的乐趣成为一种"风格"，人们开始消费自己，忘记了什么有意义以及是什么赋予了他们生活的意义，那就是挑战。

　　就建筑遗产这一概念，直到20世纪60年代，对具有纪念意义的建筑的维持和修复才成为迫切的全球问题，因为它们在战争和冲突中被摧毁了[1]。但并不是所有的建筑都能被认为具有纪念意义，也并不是所有的简单整修的修缮行为都是耗尽精力的。在教科文组织颁布的一系列纲领中有此类的文献记载，世界才逐渐认识到文化遗产的概念包括很多的要素，包括不同的文化，无形的人工产物（例如，音乐）、居

All over the world, people flourish and leave their trace. Such evidence survives in the form of tangible and intangible artefacts, part of which are the very houses people built to inhabit their land. Over the ages these houses receive various uses, modifications and even strain – from natural forces, struggle and war, or random disasters. Nevertheless, a significant part of them survives the passing of time to stand as testaments of their past, sometimes in the form of ruin and sometimes well preserved and carrying on.

The way we treat this heritage is part of an ongoing discussion, founded on the principle that no culture can explore its future by forfeiting its past. This makes sense once we realize how the technological advances of our industrial revolution gave us the power to consume our very cultural and natural environment at an unprecedented rate, and to stand brave against the inevitabilities we once were obliged to suffer (natural disasters, maladies, or man's plain limitations against his own habitat). In a nutshell, we are now technologically empowered to decide whether we comply with what was once given, or to break with it abruptly. Insofar as dwelling is concerned, we can build whatever we want wherever we want, modifying the natural landscape according to our will, need, and desire, employing materials and techniques that are immune to the restrictions of a particular location. We can dwell according to our wishes.

The question then becomes not "how", but "why". Why do we choose to build this way or the other? Why would we choose this plot of land, or that place in the world? Why would we choose to build in this style or the other? Whenever people enjoy this level of freedom, they usually end up facing their own vanity. Aesthetic pleasure becomes "style" for statement's sake, and people consume themselves, forgetting what was once meaningful and giving and what once gave purpose to their lives: a challenge.

As far as the notion of built heritage is concerned, it wasn't up until the 1960's that the call for the preservation and rehabilitation of monuments emerged as a pressing global issue, in view of the possibility of their destruction in times of war and

瑞士Scaiano旧石屋改造项目
Old Stone House Conversion in Scaiano, Switzerland

住的概念以及不断积累的记忆²。在这种背景下，我们能够认识到再利用的真正重要性：历史符号的保存比单纯的博物馆陈列要重要得多，建筑师必须将新鲜的生命注入到这些历史当中，赋予它们走向未来的力量。

同样，我们处理多余建筑物的方式也被认为是一个广泛的跨学科的调查。一方面，我们所使用的方法包括对它们现存状态进行仔细的文献记载；我们也要理解在历史背景下它们的地位，也就是在当时环境中它们的角色，包括当时的社会、文化传统、历史遗产的环境或自然景观；另一方面，我们还要让它们适应当代的要求和功能，反映新的生活水平。因此，它们可能会接受大量的干预工作，这些干预最后主宰着对话——历史和创新之间的矛盾争论；过去的魅力和未来的吸引力之间陷入的僵局谈判。

我们的社会有希望从这样的争论中解放出来。当今，革新的概念不仅仅反映出现代理想的一些东西，这些理想在盲目的技术驱动的单调中耗尽了自己，这些单调的东西颂扬现在的成绩，否定了对历史先例的需求。相应地，历史不应该被当成人类进化史的直线记录，因人类进化史需要盲目地依附学术性的真实事件而得以前进。相反地，我们现在吸取更多的概念性的东西，例如"可持续性"、"公众性"和"人文主义"价值观，"叙事主义和个性"，看起来仿佛我们试图探求抽象价值而不是具体灵感之间的新的平衡。实际上，我们不是试图重新设计我们的居住地，看起来我们更像是重新定义我们的生活习惯。从这个观点来说，试图与充满历史情感的遗址毗邻而居，说明了这一观点，即文化遗产是活的不断成长的，继而取代了有关记忆的概念，这种记忆需要用甲醛等化学物质保护起来，不受侵蚀。这是一种人和建筑之间的"位置生态学"，因为它存在于记忆、土地使用和现代属性之间。

另一方面，一旦我们不把"居住"当成一种功能，而是一种个人历

conflict¹. But not all buildings can be deemed as monuments, nor can all acts of preservation be exhausted in mere restoration. As documented by the successive charters of UNESCO, the world gradually came to acknowledge that the notion of heritage should embrace a larger variety of candidates, including diverse cultures, intangible artefacts (e.g. music), the notion of a living, and ever-growing collective memory². In this context we also can recognize the true importance of reuse; the tokens of our past are far too important to be preserved as mere museological exhibits. It is imperative to be able to breathe new life into them in order to empower them towards the future.

Similarly, the way we approach redundant buildings has been recognized as a case of extensive multidisciplinary research. For one thing, the methods we use necessarily involve the rigorous documentation of their existing condition, as well as understanding their place within their historic context equally to their role within their immediate environments – including contemporary society, cultural tradition, heritage settlements, or natural landscape. At the same time, we also need to make them adapt to contemporary requirements and functions, reflecting new standards of living. Therefore, they are expected to receive a substantial amount of interventions, which can ultimately steer the conversation of a contradicting argument between history and novelty, a stalemate quarrel between the allure of the past and the appeal of the future. Hopefully our societies are already emancipated from such arguments. Nowadays the notion of novelty reflects something far from the modern ideal, which exhausted itself in a blind, technologically driven uniformity that glorifies the achievements of the present and negates the need for historic precedent. Correspondingly, history is rarely perceived as a linear record of human evolution that needs to be preserved with blind academic adherence to authenticity. On the contrary, we now seem to inquire more upon notions such as "sustainability", "communal" and "humanistic" values, "narrative and identity", looking as if we seek to explore a new equilibrium between abstract values, rather than concrete aspirations. In fact, instead of trying to redesign our habitat, it looks as if we are aiming to redefine our habitus. From that point of view,

奥地利斯泰尔花园住宅
Garden House in Steyr, Austria

史,这些住所就会成为一个重写本,一幅包含历史的画卷,在原有的历史基础上书写,并期望被未来的历史取代。突然我们变得很谦逊,我们变得很大胆。我们的选择中存在着我们对可持续性发展的深刻理解。我们选择拥抱我们自己或者别人的过去,我们选择尊重墙上的锈迹,我们选择压抑我们自己强烈的渴望。我们的决定远远超过了功能性和必要性,它们几乎重新创造了当地的传统和条件,拓宽了未来的可能性,也拓宽了被认为建筑行为的参照系统。

"居住在历史和现代"之间的混合类型也涉及科学和技术之间的立场。历史遗留的建筑外壳作为对旧材料以及过去空间和建筑技术安排的证明,反映了旧时所有的决定都要服务于必要性和限制性。简单地说,这些建筑的身上体现着受自然条件以及特殊历史时期和产地的社会规则限制的建筑痕迹;建筑本身也反映出历史不同的层次,这些主要通过对原有建筑的连续修缮和加建体现出来。再一次,简单的专注于功能性的行为,例如,结构的一致性和适宜性,成为文献分类和防护措施范围内的问题。我们的技术,也就是我们发明和应用的才能和履历,就像是一个老大哥,一个聪明的监护人,不仅保护着年轻的兄弟,而且让他们有力量以平等的新形象屹立起来。游走在记忆、技术和美学之间的三重方法再一次占据了中心舞台。当修复的过程得到建筑学上的认可,历史的踪迹就会成为远比简单的装饰更有价值的东西。

奇怪的是,新的全球性建筑不断阐释着我们的主张。我们需要旧的建筑和新的建筑比肩而立来表达当代的条件。仿佛这些建筑在说"我们不代表未来,也不代表过去,但是我们骄傲地面对我们为自己所做的决定。"这是个有希望的消息。由于当前的建筑体系不能提供封闭的系统(就像案例中因为学术目的对纪念碑进行修缮),我们最后还是要依赖我们行业普遍的价值观,例如,功能性和美学性。就像建筑语言所陈述的,我们可以将我们的设计提议与折中主义、极简主义、高

choosing to live in juxtaposition to historically charged sites exemplifies the notion of a heritage that is living and growing, replacing the idea of a memory that needs to be preserved in formaldehyde. It is a sort of "situated ecology", between man and building, inasmuch as it is between memory, land (re)use and modern commodities.

On the other hand, should we consider "dwelling" not as a function but as a personal history, these dwellings become a palimpsest, a scroll that contains histories, that is written upon previous histories, and that is expected to be substituted by future histories. Suddenly we become modest. And we become bold. In our choices lies a deep understanding for continuity. We choose to embrace our own or someone else's past, we choose to respect the patina on the wall, and we choose to stand tall against our very aspirations. Our decisions go beyond mere functionality or necessity: as much as they reinvent local traditions and conditions, they also broaden the range of possible futures and widen our systems of reference for what may be deemed an act of architecture.

This hybrid type of "dwelling in-between" also involves a position between technology and technique. The old shells stand as testaments to materials, arrangements of space and building techniques of the past, reflecting a time when decisions were bound to a system of necessities and restrictions. To put it simply, they carry an exact image of the natural limitations and the rules of society for their specific time and place of origin – and as such, they can also reflect various layers of history manifested through sequential modifications or additions to the original construction. Once again, simple functional preoccupations such as structural coherence and adequacy become questions of documentation and safeguarding. Our technology, namely our capacity and record for invention and application, must act as an older brother, a wise guardian who will not only protect but also empower the youngest sibling to stand anew as an equal. The threefold approach between memory, technology and aesthetics takes, once again, center stage: when the rehabilitation process becomes recognizable as architecture, the traces of our past become something far more valuable than plain decoration.

Oddly enough, new and global architectural expressions inform

西班牙赫罗纳奥洛特某联排住宅
Row House in Olot, Girona, Spain

科技或者简单的美学优雅等目的结合起来。我们可以在模仿的策略中做出选择;因为模仿,我们可以挪用旧建筑的结构和特质;因为模仿,我们通过新旧建筑彼此之间的矛盾和融合在它们之间建立对话,这样旧的和新的建筑就会融合成不可分割的整体。

我们在这方面研究的项目是上述方法典型的代表。然而,最重要的是,它们刻意表达一种多样性来鼓励我们克服胆怯。这些建筑向我们展示了什么在技术上是可能的,什么在功能上是可行的,什么在美学上是合适的。这些建筑向我们的行业展示出新的可以研究、可以实施我们想法的领域。这些领域与新的自由先进的建筑结构相比,与公众的沟通较少。在我们寻求出这些可能性的时候,这些案例就向我们展示了历史遗迹如何充满新生活的元素,进而将建筑更好地传递给未来的一代。

由Irisarri-piñera S.L.P.建筑事务所设计的位于黛波拉的独栋住宅,建立在一个古老的当地农舍的废墟基础上。具有奇怪形状和大体量的全新白色石质建筑醒目地屹立着,且其沿着原有的建筑结构轮廓而建的空间、自然风景的斜坡和为了适应家庭居住需求而设计的空间使主要的建筑思想展现出一系列过渡性的特点。与通常的过渡空间设计一样,这些空间都经过有形和无形的薄膜过滤,薄膜能调节空气、光线、人的行动与感知。时间看起来并不是静止的,而是随着建筑空间的改变而变化的。

由RCR建筑师事务所建造的奥洛特某联排住宅与其说是一座混合建筑,不如说是一种建筑思想的声明。因为政府对沿街建筑立面的修缮有限制,建筑师只能在大空间的质量上做文章,这个空间策略性地安排在两座相邻建筑原有墙体之间,空间面积有限,因此只能在高度上合理安排功能区。这个空间为这座像镜子一样的建筑打下了基础,显示了整座建筑的本质:身处屋内,你可以透过两层高的玻璃立面看

our propositions relentlessly. We need both the old and the new to stand together equally in order to express a contemporary condition. It is as if these dwellings say "we are not the future, nor the past, but we stand proudly upon the decisions we made for ourselves in the present". That is a hopeful message. Since the present tense cannot afford to be a closed system (as the case would have it with an academically driven restoration of a monument) we end up relying on the universal values of our trade, such as function and aesthetic beauty. As the language of architecture would have it, we can apply design proposals driven by eclecticism, minimalism, hi-tech, or just plain aesthetic elegance. We can choose between strategies such as mimesis, out of which we would appropriate the older structure, distinction, out of which we would establish a dialogue between the old and the new through contradiction, or confluence, which would seek to merge the old and the new into an indivisible whole.

The projects we examine in this issue are exemplary to the aforementioned approaches. Yet, and most importantly so, they intentionally express a variety which seeks to embolden us against timidity. They show us what's technically possible, what's functionally feasible, and what's aesthetically pleasurable. They present us with a field for our trade, to research or to act upon, which is often poorly communicated to the general public in comparison to new, freely developed architectural constructions. As we seek out the possibilities, these case studies show us how our relics can be invested with new life in order to be passed onto future generations.

Single Family House in Tebra by Irisarri-piñera S.L.P. is developed upon the ruins of an old vernacular farming house. While the new insertions stand out distinctively in their white color, odd shapes and volumes, the principal idea unfolds a series of transitions through space, following the contours of the pre-existing structure, the slopes of the natural landscape and the spaces that are tailored to the family's living requirements. As usual with transitions, this array of experiences is filtered by tangible or intangible membranes that modulate air, light, movement and perception. Far from standing still, time seems to follow the architectural pace.

Row House in Olot by RCR Arquitectes is far less a mix than it

到院子，并且面对周围的自然环境、生动的建筑和广阔的天空陷入沉思。当你来到室外，建筑内部的全部景象就会马上戏剧性地展示在你面前，笼罩在极简抽象风格的光芒中。

  由Hertl.建筑事务所设计的花园住宅醒目地矗立在毁掉的旧农舍之上以及内部。两座建筑的材料都是未经处理加工的：裸露的石质结构、混凝土结构和自然风化的木板。这种混合的结构贯穿在整个空间及体量的连接之中。Hertl.建筑事务所几乎没有依据旧建筑的轮廓建造新建筑，而是巧妙地处理空隙空间，甚至在较宽的地方另建了一座"相似的建筑"。两座建筑的功能区被建筑师们划分为三个区域，Refugium区被用作休闲后院，Laboratorium区作为创意工作坊、开会演讲以及讨论之用，而Klausur区则作为私人观赏与展览会和音乐会等小型文化活动的举办场地。

  最后，由Wespi de Meuron Romeo建筑师事务所负责的Scaiano旧石屋改造项目是一个非常典型的例子。该事务所非常善于处理建筑的自然元素和古老特质的结合。建筑被嵌入一座经时间洗礼的壳体中，壳体既可以作为旅馆也可以作为帐篷。新建筑仿佛本来就伫立在那里，没有违和感；那些"奇怪的细节"显示出它是一座充满诗意的后建建筑，来自于一个时光静止的时代。光、空气和温度等基本要素营造出远比功能性设计更好的空间氛围。最终建成的建筑是集沉静与凝重、独特与普通、复杂与简单为一体的建筑，仿佛它直接由时光与记忆雕刻而成。

1. UNESCO World Heritage Center, "The World Heritage Convention / Brief History," accessed November 6, 2015, http://whc.unesco.org/en/convention/.
2. UNESCO World Heritage Center, "The Criteria for Selection," accessed January 6, 2015, http://whc.unesco.org/en/criteria/.

is a statement. As the limitations imposed by the authorities concerned only the restoration of the main facade on the street, the architects were left free to ponder upon the quality of an imposing void space strategically placed in-between the pre-existing walls of the adjacent properties, arranging the functional zones of the house at height in their bare dimensions. This void sets the stage for a mirror-like spectacle which reveals the essential nature of things: once you are inside, you can look through a huge two-storey glass facade towards the courtyard and meditate into the surrounding nature, the graphic buildings and the open sky; yet once you are outside, the house theatrically presents its whole interior at once, in all of its minimalistic glory.

The Garden House by Hertl. Architekten ZT GmbH rises distinctively on top, and from within, a ruined old farmhouse. The elements of both structures are left raw: the uncovered stone structure meets with concrete volumes and naturally stained wooden planks. The idea of a hybrid continues onto the articulation of spaces and volumes. Hertl. Architekten ZT GmbH hardly follows the contours of the old structure but rather plays with it by manipulating void space and seeking to "counterpart" buildings on the wider area. The functional content of the ensemble is described by the architects as "Refugium for retreating, Laboratorium for creative workshops, lectures and discussions, and Klausur for private viewings and small cultural events such as vernissages and concerts". Finally, the Old Stone House Conversion in Scaiano by Wespi de Meuron Romeo Architects is exemplary to the firm's masterful articulation of the elemental and archaic quality of architecture, embedded into a time-stained shell that serves both as a host and as a canvas. The new building stands as if it has always been in place, while its "strange details" reveal its poetic descendance from a place where time stands suspended. Fundamental properties such as light, air and climate create space in a far more efficient way than mere functionality. The end result is composed and dense, unique and ubiquitous, intricate and simple, as if it was carved out of time and memory directly. Angelos Psilopoulos

奥洛特某联排住宅
RCR Arquitectes

我们清空了一座立面为列管建筑遗产的联排住宅内部，释放了由这些古老的墙体和屋顶封闭的空间。这是一个独立的空间，由高度不同漂浮式平台连接起来。其中几个平台可作为长凳，既保护了空间又界定了空间。建筑结构占据场地整个长度，与两侧墙体分离的一个过滤空间有助于平台之间的交流。

该住宅可以根据主人要求而选择打开或者关闭与自然景色的视觉联系。花园的景色也在不断变化。住宅后面有一座附属建筑，围合出一个开放的空间，使人们在整座建筑中都能感受到这个空间的存在，交叉体验室内外景色的同时又不失私密性。

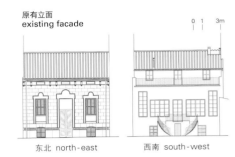

项目名称：Row House
地点：Olot, Girona, Spain
建筑师：RCR Arquitectes
项目团队：A.Lippmann, R.Muñoz de León, Blázquez-Guanter, arquitectes, Artec3 Studio, D.Leikina, X.Loureiro, M.Houille
用途：single-family dwelling
占地面积：383m²
设计时间：2009 / 施工时间：2010 / 竣工时间：2012
摄影师：©Eugeni Pons

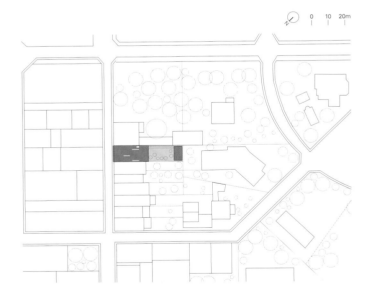

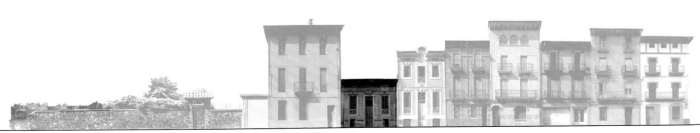

东北立面 north-east elevation

Row House in Olot

We empty a row house with a classified heritage facade, and release the air enclosed by these ancient walls and their roof. This is a single space, articulated by floating platforms set at different levels. Several of the platforms serve as seating benches that protect and frame the space. The structure runs the full length, and a filter separated from the walls on either side organizes the communication between the platforms. The house section turns on and off the visual relations of scene that changes with the personalities. The frame of the garden also varies. An annex at the rear encloses a world that is open to the air and the sky, allowing it to be sensed throughout the house while crossed experience is shared without a loss of privacy. RCR Arquitectes

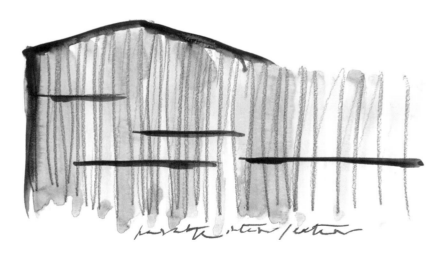

142

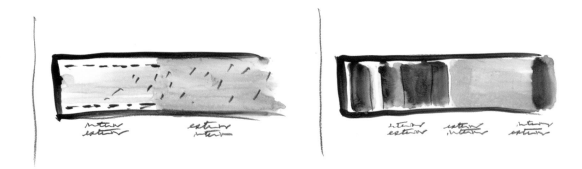

平台层 deck level

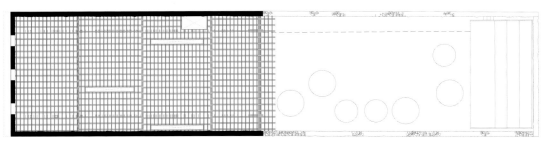

1.私密区域楼梯
2.卧室
3.机械室楼梯

1. stairs in private area
2. bedroom
3. stairs in machine room

上层 upper level

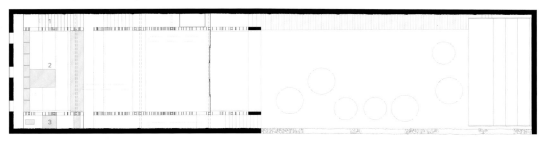

1.入口
2.公共浴室
3.公共区域坡道
4.厨房
5.烟囱
6.起居室
7.室外平台
8.花园楼梯
9.私密区域楼梯

1. entry
2. common bathroom
3. ramp in common area
4. kitchen
5. chimney
6. living room
7. outdoor gallery
8. garden stairs
9. stairs in private area

入口层 access level

1.花园楼梯	5.游泳池	9.浴室	13.亭子台阶
2.私密区域楼梯	6.水池	10.洗浴间	14.阶梯式亭子
3.设备区	7.机械室	11.淋浴间	15.花园
4.主卧	8.机械室楼梯	12.书店	

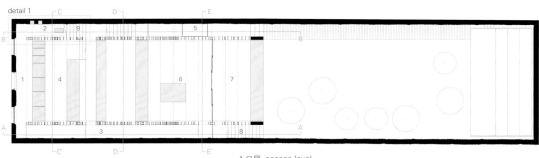

1. garden stairs
2. stairs in private area
3. service area
4. master bedroom
5. swimming pool
6. sheet of water
7. machine room
8. stairs in machine room
9. bathroom
10. bathing pavilion
11. shower
12. bookstore
13. stairs in pavilion
14. step pavilion
15. garden

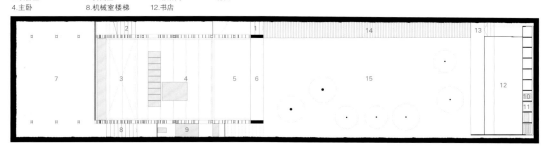

下层 lower level

0 2 5m

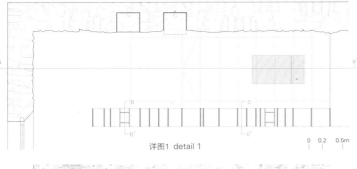

详图1 detail 1　　　　0　0.2　0.5m

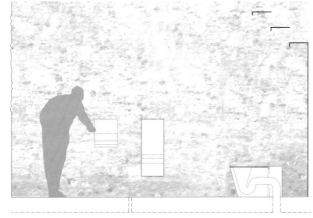

a-a' 剖面图 section a-a'　　　　b-b' 剖面图 section b-b'　　　　c-c' 剖面图 section c-c'

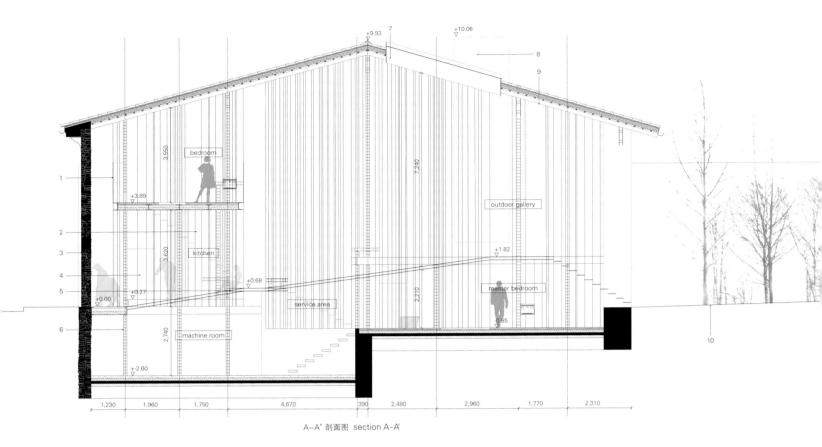

A-A' 剖面图 section A-A'

1. walls: steel sheet(e=8mm)
2. stairs: steps stainless steel plate S275JR Jindal (e=15mm), welded to vertical surface
3. facade: existing facade, towards the outside and inside stonework
4. furniture: kitchen furniture stainless steel sheet
   S275JR Jindal (e=5mm)
5. ramp: tubular profile, with ceramic tile and concrete layer 6cm,
   interior and top finish with stainless steel sheet 6,500x1,000x5mm
   structure: pillars composed of two steel S275JR (e=8mm)
6. skylight: edge formed by black steel sheet and laminated glass with air chamber 6+6
7. upper sun protection with roller blind
   chimney: folded steel sheet S275JR jindal (e=8mm)
8. stairs: 5mm plate with 70mm bottom rib
9. embedded in masonry wall
10. garden: topsoil
11. structure: reinforced concrete diaphragm wall with volcanic aggregates 100x25cm
12. existing wall: volcanic stone masonry with plaster (e = 40cm)
13. walls: 5mm plate (h=150cm)
14. stairs: steel sheet S275JR (e=5mm)
15. garden: path on grass, steel handrail
16. cover: existing wooden beams(210x80), support girder,
    wood battens 70x35cm
    breathable waterproof, cork insulation 50mm
    concrete made with arlita
    mortar and ceramic tiles
17. stairs: concrete stairs into the garden
18. structure: wrought steel deck on beams HEB200
    steel plate S275JR Jindal (e=5mm)
    self leveling (e=2mm), concrete protection (e=2.5mm)
    polystyrene support
    insulation (e=10mm), tubes (ø16mm)
    deck plate (e=6cm), rock wool insulation (e=10cm)
    metal sheet (2mm)

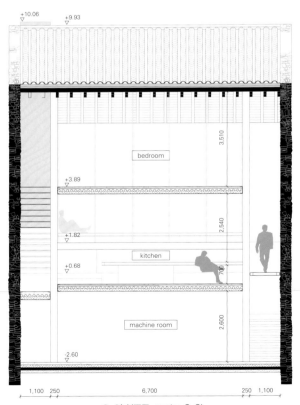

C-C' 剖面图 section C-C'

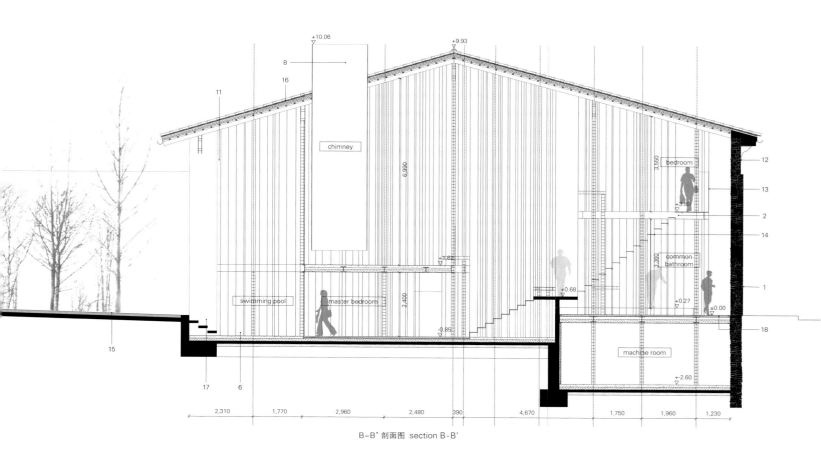

B-B' 剖面图 section B-B'

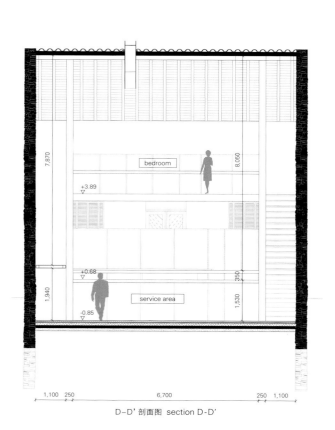

D-D' 剖面图 section D-D'

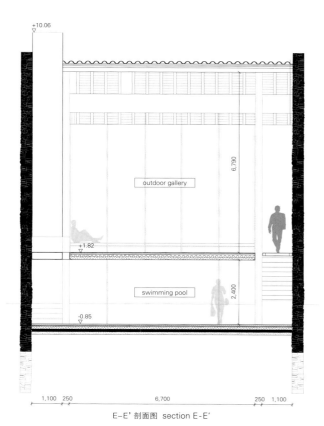

E-E' 剖面图 section E-E'

黛波拉某独栋住宅

Irisarri-piñera S.L.P.

在这座建筑中，你会享受到漫步、居住和捕捉自然的美妙时刻。这个古老而传统的农舍结构使用了一系列原来自然存在的岩石；掩映在繁茂枝叶下的桌子见证了无数次正餐和几代人的家庭聚会；住宅附近还有一个水池。所有的这些元素成为设计的灵感来源，这里也必将成为一个生活场所。此外，使这里成为一个生活场所的还有下列元素：装饰有小台阶的斜坡，能够避开废墟的高耸的赤褐色岩石，这块土地上的生活经验，一个面向西南努力向着太阳的山谷，夏日漫长岁月集聚的热量和晚秋对光的渴望。所有这些，都成为每日微小的精彩瞬间。

根据经验，建筑师通过创造连续的中间空间实现了和废墟（三个排成一排的废墟）的连接，将废墟不同的情况和确定它们位置朝向的特点统一起来，使这个项目除了拥有原来提议的新建空间外，还拥有了更多的补充空间。

通过这种方式，我们探索了这些空间进行气候、光照或私密性控制的不同形式。这些方式作为一个系统，有效地实现了景观和那些美妙时刻的连接。

最私密的区域设置在建筑的两端，是一些独立的空间。在建筑这两端之间，我们顺应地势，建造了连续的封闭空间。

先前废墟具有非常含蓄的特征，有利于形成内部的私密亲密空间。这些空间与重建的外部区域以及墙体和废墟形成的开放庭院形成鲜明的对比。

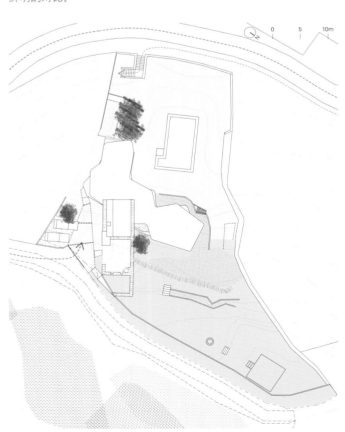

原有建筑的残留部分被当作废墟来处理，其砌石墙体通过覆盖混凝土来进行加固。连续的混凝土板能让人想起建筑原始的体量，进而强调了古老建筑的私密特性。

混凝土板模仿了这个地区的地形特征，界定了连续的空间。建筑师用了一种特殊的分层稍微封闭一下，起到主动隔膜的作用。这种混凝土板朝向不同，构成也不同，可操作性很强，使用者也可以对其进行进一步处理。

密封层和混凝土板的错位安装，共同按居住者的意愿营造出朝花夕拾的生活空间。

Single Family House in Tebra

Wandering, living, capturing moments. A series of pre-existences, petrous rests of an old traditional farming structure, a table under the leafy glycine, that has greeted meals and family reunions throughout decades, and the nearby pool, all become the origin and the plot that set the space bound to be a living place. Also, a sloping terrain garnished with small terraces, an auburn that grew up sheltered from the ruins and the experience of life in the parcel, the sun's haul at a southwest-facing valley, the heat in those long days of summer, and the wish of light for the late fall become every-day little moments.

It is a matter of experiencing through the generation of a continuum of intermediate spaces newly built that link the ruins, three pieces grouped in line, and bringing together their diverse conditions and characters that give them their position-

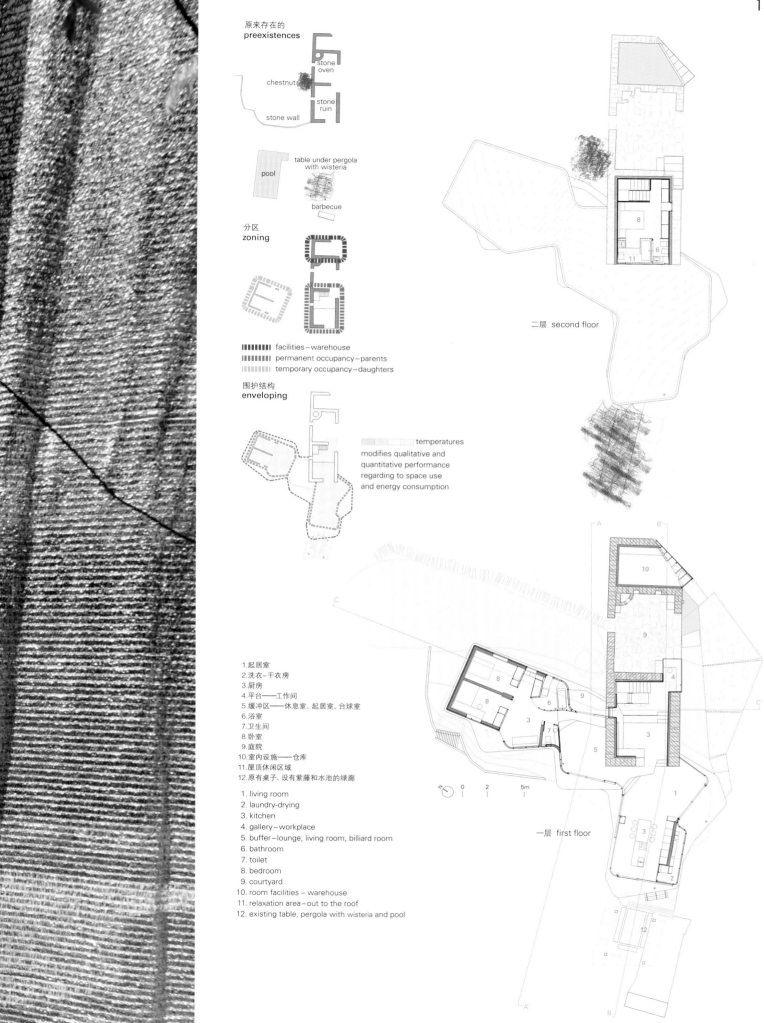

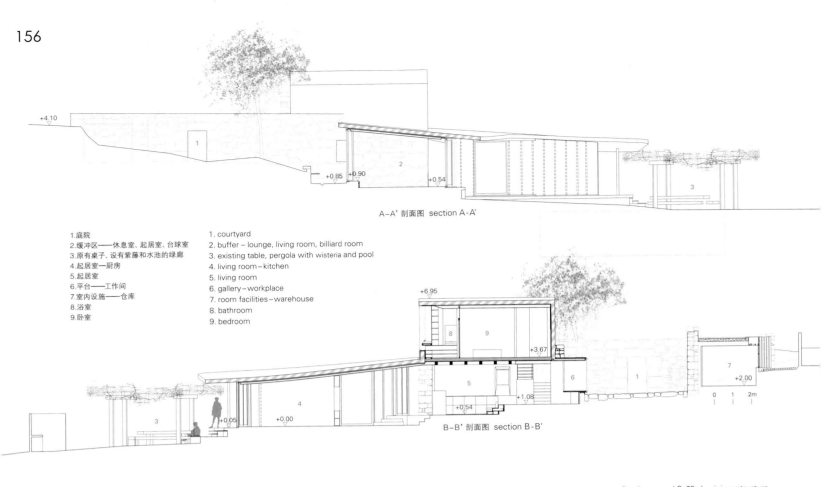

A-A' 剖面图 section A-A'

1. 庭院	1. courtyard
2. 缓冲区——休息室、起居室、台球室	2. buffer – lounge, living room, billiard room
3. 原有桌子、设有紫藤和水池的绿廊	3. existing table, pergola with wisteria and pool
4. 起居室—厨房	4. living room – kitchen
5. 起居室	5. living room
6. 平台——工作间	6. gallery – workplace
7. 室内设施——仓库	7. room facilities – warehouse
8. 浴室	8. bathroom
9. 卧室	9. bedroom

B-B' 剖面图 section B-B'

1. 20mm cellular polycarbonate sheet, colourless finish,
   attached to 60.60.15 aluminium pipes with aluminium corner brackets,
   the brackets are pierced each 20cm to lend the ventilation
   of dust-prevention strip
   multilayer reflective insulation (30mm)
   inner tube (4cm)
   reinforced concrete wall (20cm)
2. aluminet 80-O thermoreflected net fastened up to the aluminium rail
   placed in the roof raindrop
3. pre-existent double side granite wall (60cm)
   reinforced concrete wall (10/15cm) anchored to the stone wall
4. coniroof elastic waterproofing
   reinforced concrete slab (20cm)
   insulation, extruded polystyrene (6cm)
   lining ceiling, plasterboard panelling over galvanised steel structure
5. polymeric cementitious coating
   concrete slab over/under floor heating
   radiant underfloor heating and cooling system
   reinforced concrete slab 8cm thick, over polypropylene blocks
   poor concrete slab / compacted soil
6. polymeric cementitious coating
   radiant underfloor heating and cooling system
   reinforced concrete slab (10cm) / damp-proof plywood (25mm)
   conifer serrated beam
7. concrete and boulders slab / elastic waterproofing
   concrete slab with slope
   reinforced concrete base
8. schuco royal S-65 aluminium structure
   planilux 5+5/15/4+4 glass
9. dry wall construction 12.70cm thick
   plasterboard panelling (13+13mm)
   galvanized steel structure 7.5cm thick
   insulation, mineral wool
   plasterboard panelling (13+13mm)

C-C' 剖面图 section C-C'

夏季白天 summer day

夏季夜晚 summer night

1. buffer-lounge opened as a porch
2. cool breezes from the courtyard shaded by chestnut tree and the pool water through the porch
3. cross ventilation
4. thermo-reflector aluminized screens generating intermediate shadow spaces
5. solar panels for hot water
6. dissipation by natural convection in ventilated facades
7. folded thermo-reflector aluminized screens

冬季白天 winter day

冬季夜晚 winter night

1. heat accumulating greenhouse
2. thermal mass
3. hot-cold underfloor
4. folded thermo-reflector aluminized screens
5. solar panels
6. extended thermo-reflector screens as a barrier to the exchange of temperature by convection

orientation, to provide the project with complementary areas in addition to those new ones previously proposed.

In this way we explore the manners that generate different sets of filter control: climate, light or privacy; as a system of joints, it connect landscapes and moments.

The most private areas are disposed with some independence at the ends and between them a continuous space wraps adapting to the topography.

The previous character of the ruins' spaces, more introverted, contributes to the domestic area with those most intimate moments that contrasts with the exterior in the reconstructed region, as well as open courtyards where the walls and ruin had generated them.

The remains of the existing building are treated as ruin by consolidating the masonry walls by cladding concrete that continues as slab evoking the original volume, and thus emphasizing the private nature of the ancient construction.

A concrete slab that mimics the topography of the area and defines the continuous space, with a slight enclosure of specialized layers that performs as an active membrane, with diverse compositions depending on their orientation and high capacity for handling and further processing by the users.

The offset both of the sealing layers and of the slab creates ephemeral spaces at the will of the inhabitants.

项目名称：Single family house in Tebra
地点：O Casal, San Salvador de Tebra, Tomiño(Pontevedra), Spain
建筑师：Irisarri-piñera S.L.P.
结构工程师：Ibinco S.L.
机械工程师：Exinor S.L.
工料测量师：Manuel Taboada Acevedo
设计时间：290m²
施工时间：EUR 260,622
竣工时间：2006
竞赛时间：2011
摄影师：©Hector Santos Díez

Scaiano旧石屋改造项目
Wespi de Meuron Romeo Architects

这座历史悠久的石屋位于Scaiano村的中心,原来被包含在一座主楼内,主楼设有带拱顶的酒窖,厨房、起居室和卧室位于上面两层。后来新建的附属建筑也完全由坚固的天然石材建造而成,一层包括一个小型的葡萄白兰地酒厂,二层有一个房间,上面还有一个阁楼。室外石阶和主立面前面的几段石墙似乎都是与附属建筑在同一时间建成的。主楼的前面原来是一个小广场,在建造附属建筑之前,它是村里建筑中唯一的一个广场。

建筑在改造之前主人在此处已居住了二十多年。全凭屋顶的保护,它才能延续至今,但为了适应现代住宅对舒适性的要求,这栋房屋需要进行一次彻底的改造。

目前,这座石屋的石墙大体上被保留了下来,只有连接处需要重新塑造。墙体内侧增加了隔离层,所有的木梁楼板都被换成了混凝土楼板,起到加固旧墙的作用。主楼石墙上方插入了一个新的木结构坡屋顶,同时附属建筑上建造了一个由天然石料铺设而成的屋顶露台。

嵌入结构旨在塑造出大型石墙的力量之美,并且使这栋历史建筑的体量看起来更加古朴简约。所以,建筑师拆除了建筑外部的石质楼梯,在房屋前面重新建造了小型村广场,增加了喷泉和长凳,人们可以在这里进行会面。

带拱顶的旧酒窖将成为建筑新的主入口,它被分为没有暖气的室外区域和有暖气的入口处两个区域,入口处设置有一个衣柜和一个带有洗手池和淋浴间的小房间。一旁的石墙被切割掉一块,建造了通往楼上区域的楼梯。旧的白兰地葡萄酒厂现在也与拱顶酒窖连接,夏天可以作为凉廊,内设简易的粗钢壁炉。

房间的正面采用玻璃立面,立面面向湖水,与旧石墙的距离大约为60cm。这面玻璃墙就像立面后面的第二层皮肤一样,保护着内部免受室外天气的干扰。这个室外区域没有屋顶,并不能够挡住风雨。

设计理念一方面真正地保护了能够叙述这栋房屋历史的石头立面,另一方面,通过特殊的光反射方式,也为房间创造了最大限度的阳光照射量。否则在街道如此狭窄的村落,是不太可能有如此大量的阳光进入房间的。

室内采用了与原来古色古香的环境匹配的设计:经过天然浸染的水泥地板、石灰色抹灰的墙体和天花板、刷油的落叶松木家具。新的设计并没有与原有的设计形成对比,而是新与旧相互融合,形成了一种新的整体感。

Old Stone House Conversion in Scaiano

The original substance of this historic stone house, in the core of the village of Scaiano, consists in a main building with a cellar with vault and two floors above with kitchen, living and sleeping spaces. A later added annex, also completely made in solid natural stone, contained on the ground floor a small grape brandy distillery, on the first floor a room with above an attic. Exterior stone stairs and some small walls in front of the main facade, seem to have been created in the same time as the annex. In front of the main building there was originally – before the annex was built – a small square, the only one in the village structure.

西北立面 north-west elevation   西南立面 south-west elevation

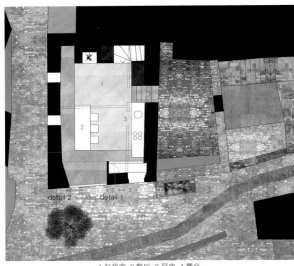

1.起居室 2.餐厅 3.厨房 4.露台
1. living space 2. dining 3. kitchen 4. terrace
三层 second floor

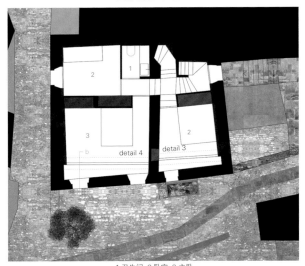

1.卫生间 2.卧室 3.主卧
1. toilet 2. bedroom 3. master bedroom
二层 first floor

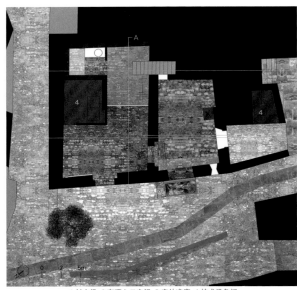

1.村广场 2.有顶入口空间 3.室外凉廊 4.技术设备间
5.入口&衣柜 6.淋浴间 7.室外庭院
1. village square 2. covered entrance space 3. outside loggia
4. technique 5. entrance & wardrobe 6. shower 7. outside court
一层 ground floor

Before the conversion this building was inhabited for over two decades. It was protected with a provisory roof against decay, but it needed a complete renovation to be adapted to the contemporary needs of comfort.
The stone walls were mainly maintained, just the joints had to be remade. The isolation is applied at the inside of the walls. All wooden beam floors were replaced with concrete floors, which reinforce the old walls additionally. On the main house, a new pitched roof in wood construction was inserted in the stone walls, meanwhile on the annex it was created a roof terrace with a natural stone-paved floor.

The main architectural target of the intervention was to carve out the force of the massive stonewalls and to gain this archaic simplicity of the volume of the historic building. So the external stone staircase was removed to create again the small village square in front of the house, with a fountain and a bench, which invites to spontaneous encounters.
The old cellar with vault becomes the new main entrance: it was divided in an unheated outdoor zone and an heated entrance zone with the wardrobe and a small room with washbasin and shower. On one side the old stonewall was cut out and a stair was integrated, to lead up to the upper floors.

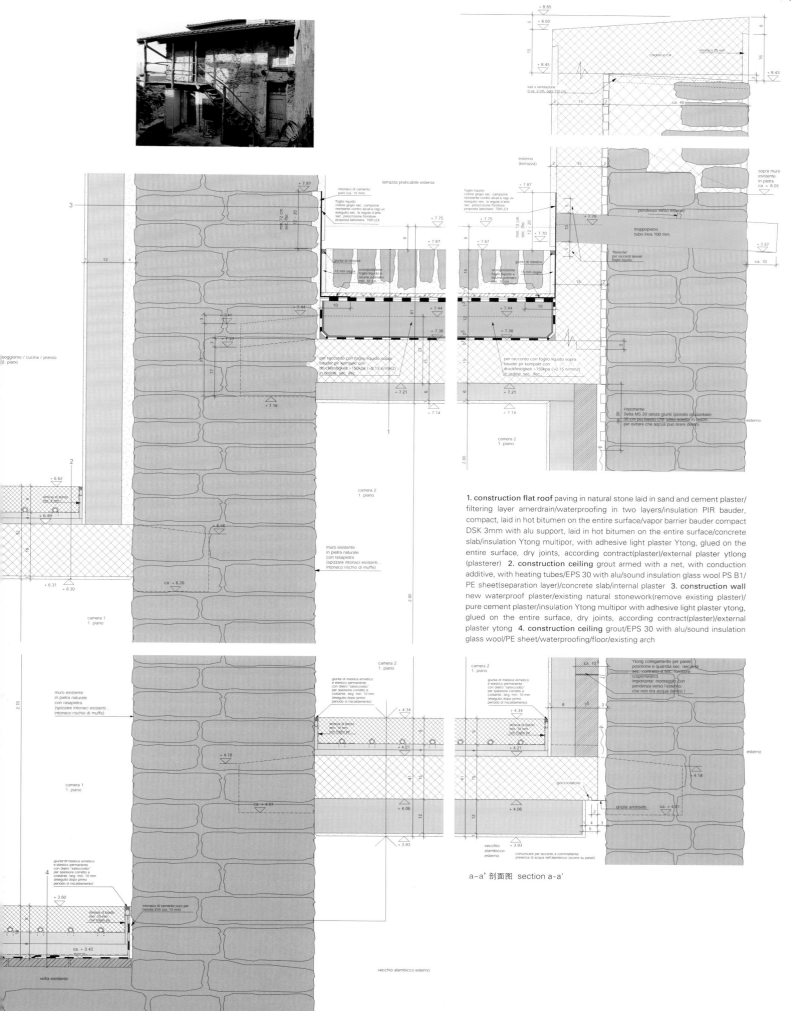

1. **construction flat roof** paving in natural stone laid in sand and cement plaster/filtering layer amerdrain/waterproofing in two layers/insulation PIR bauder, compact, laid in hot bitumen on the entire surface/vapor barrier bauder compact DSK 3mm with alu support, laid in hot bitumen on the entire surface/concrete slab/insulation Ytong multipor, with adhesive light plaster Ytong, glued on the entire surface, dry joints, according contract(plaster)/external plaster ytlong (plasterer) 2. **construction ceiling** grout armed with a net, with conduction additive, with heating tubes/EPS 30 with alu/sound insulation glass wool PS B1/PE sheet(separation layer)/concrete slab/internal plaster 3. **construction wall** new waterproof plaster/existing natural stonework(remove existing plaster)/pure cement plaster/insulation Ytong multipor with adhesive light plaster ytong, glued on the entire surface, dry joints, according contract(plaster)/external plaster ytong 4. **construction ceiling** grout/EPS 30 with alu/sound insulation glass wool/PE sheet/waterproofing/floor/existing arch

a-a' 剖面图 section a-a'

The old grape brandy distillery is now also connected to the cellar with vault and can be used as a cool summer outdoor loggia with a simple fireplace in raw steel.

The glass facade in front of the rooms, towards the lake, has been placed with a distance of about 60cm from the old stonewalls. This glass front protects against the outside climate and it's like a second skin behind the facade. This outer zone is mainly not covered and it rains in.

This concept allows on the one hand the authentic conservation of the historic stone facade, which tells the history of the house and on the other hand, it generates zenith light for the rooms with exceptional light reflections. It would not have been possible otherwise to get sunlight into the rooms, in such a village structure with narrow streets.

The materialization of the interior is adapted to the archaic existing: natural impregnated cement floors, stone gray plaster walls and ceilings and oiled larch wood joinery. The new intervention is not contrasting the existing substance, but new and old are merging together and creating a new ensemble.

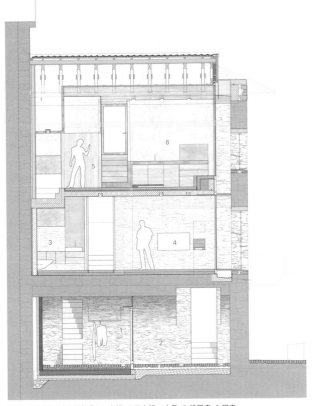

1.入口&衣柜 2.有顶入口空间 3.卫生间 4主卧 5.起居室 6.厨房
1. entrance & wardrobe 2. covered entrance space 3. toilet
4. master bedroom 5. living space 6. kitchen
A-A' 剖面图 section A-A'

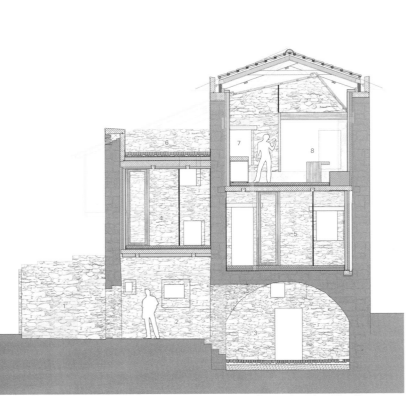

1.室外庭院 2.室外凉廊 3.有顶入口空间 4.卧室 5.主卧 6.露台 7.厨房 8.餐厅
1. outside court 2. outside loggia 3. covered entrance space
4. bedroom 5. master bedroom 6. terrace 7. kitchen 8. dining
B-B' 剖面图 section B-B'

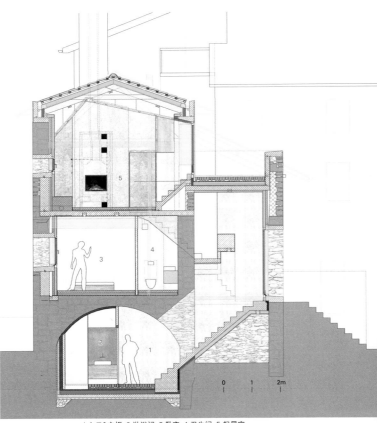

1.入口&衣柜 2.淋浴间 3.卧室 4.卫生间 5.起居室
1. entrance & wardrobe 2. shower 3. bedroom 4. toilet 5. living space
C-C' 剖面图 section C-C'

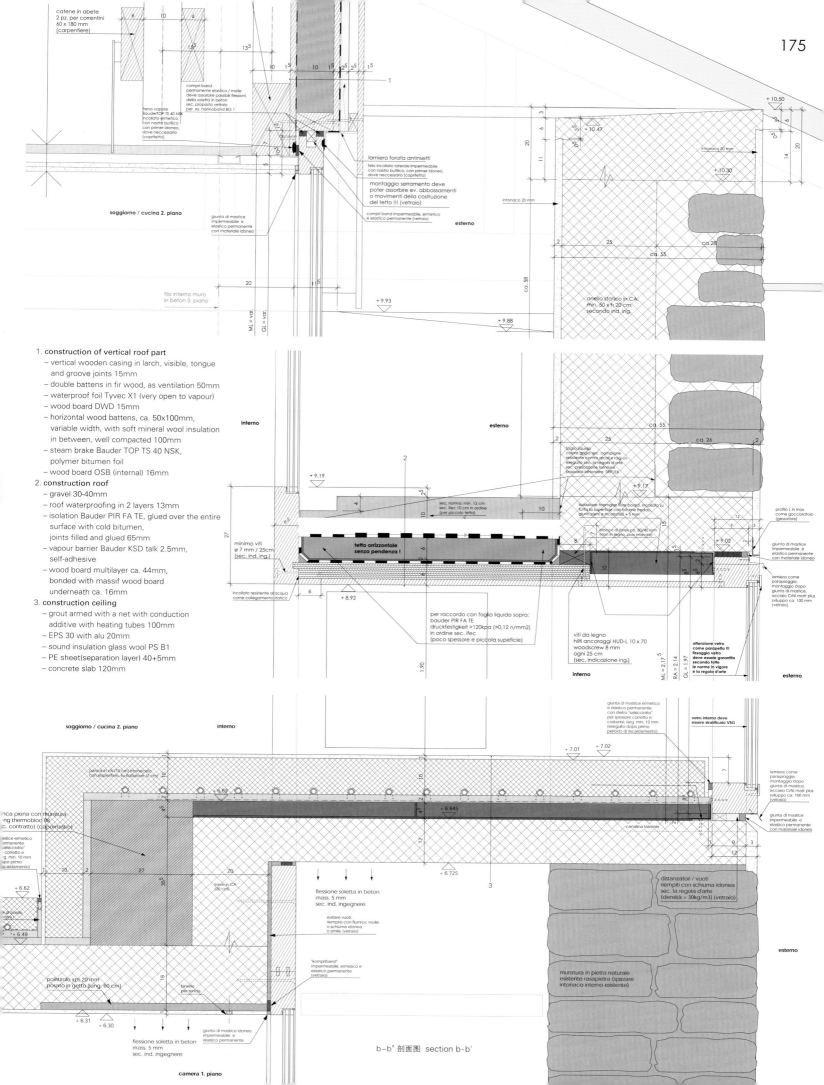

1. construction of vertical roof part
   – vertical wooden casing in larch, visible, tongue and groove joints 15mm
   – double battens in fir wood, as ventilation 50mm
   – waterproof foil Tyvec X1 (very open to vapour)
   – wood board DWD 15mm
   – horizontal wood battens, ca. 50x100mm, variable width, with soft mineral wool insulation in between, well compacted 100mm
   – steam brake Bauder TOP TS 40 NSK, polymer bitumen foil
   – wood board OSB (internal) 16mm

2. construction roof
   – gravel 30-40mm
   – roof waterproofing in 2 layers 13mm
   – isolation Bauder PIR FA TE, glued over the entire surface with cold bitumen, joints filled and glued 65mm
   – vapour barrier Bauder KSD talk 2.5mm, self-adhesive
   – wood board multilayer ca. 44mm, bonded with massif wood board underneath ca. 16mm

3. construction ceiling
   – grout armed with a net with conduction additive with heating tubes 100mm
   – EPS 30 with alu 20mm
   – sound insulation glass wool PS B1
   – PE sheet (separation layer) 40+5mm
   – concrete slab 120mm

b-b' 剖面图 section b-b'

住宅——居住在记忆里 Dwell How – Dwelling in Memory

花园住宅
Hertl. Architekten ZT GmbH

小广场在中世纪曾经是作为通往旧城区恩斯多夫的门户标志,如今这里却建成了一座花园住宅。这里曾经是一座古老的农舍,农舍遭受过彻底的毁灭性破坏,后来有了另一个身份,成为一座通行税征收所,征税所亦经历了一段悠久的历史。随着历史痕迹逐渐销声匿迹,曾经立在征税所山墙前的、给人们留下深刻印记的路边神龛,已经被迁移到了右边。此外,树篱也离这片区域太近了。

沿着Haratzmüllerstraße和凹进处狭小的轮廓而建的花园新墙体,与征税所的山墙平行排布,暗示了大型广场的存在;人们从外围望向花园住宅时只能看到茂密的树林,但能清楚看到位于广场一角的住宅入口。

场地周边的围墙让这里变得格外宁静,然而,与此同时它对那些想要走进这座现存的建筑里面观光的游客却是开放的。我们之所以称之为花园住宅,是因为其渗透出一种浓郁的绿意与空间相互蔓延交融的氛围。Refugium区被用作休闲后院,Laboratorium区被用作创意工作坊、开会演讲以及讨论之用,而Klausur区则作为私人观赏与如展览会开幕日和音乐会等小文化活动的举办场所,所有的这些阐述了该建筑的作用。

古老的农舍是一座废墟,残存的外墙围合成了一个大天井,我们在其中插入了一个新的混凝土体量,形成了"屋中屋"的设计理念。两个天井在小房子底下连接在一起形成了一个贯通的空间。

楼上(新体量)有一间客房和一个主入口。通到下面庭院的楼梯隐藏在多孔的双层新建外墙后,阳光可透过砖砌的小孔进入建筑。新体量在Enns河的方向向外悬挑,在旧墙之外创造了一个室内空间。在这里你的眼睛能够跨越河流,而这里的河岸曾经作为竹筏停泊的海港。

在节约能源、资源优化以及自然保护等方面,建筑的可持续性成为土地使用规划的一部分,它的建造基础是使用这座城市现有的结构。这可能为整个欧洲的未来给出了最佳的生态设计解答。比起其他方案,该设计避免了建造更多的新建筑,保留了更多的资源。

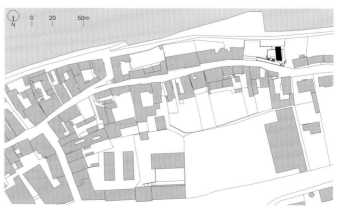

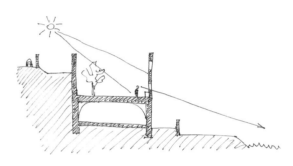

Garden House

Where in the past a small square marked the medieval gateway to the old town district of Ennsdorf, the garden house stands today. Arisen from an old farmhouse which was entirely ruinous it shapes the counterpart of the historical tollhouse. The traces of the past have slightly vanished, and the wayside shrine, that once stood impressive in front of the tollhouse's gable wall, has been moved to the right. The hedges have become too close to the free area.

The newly built garden wall follows the narrow contours of the Haratzmüllerstraße and recesses in alignment with the tollhouse's gable. Thereby the suggestion of a larger square appears and the entrance to the garden house, which one can feel from outside as lush greens only, becomes clearly visible as a simple corner of the square.

With its barriers the site ensures calmness but at the same

南立面 south elevation

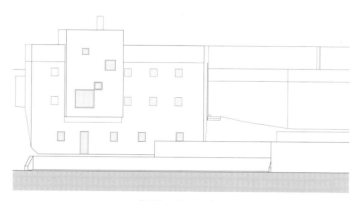

北立面 north elevation

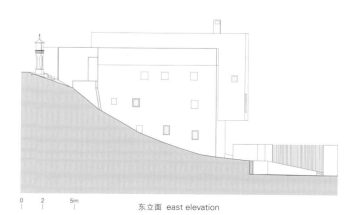

东立面 east elevation

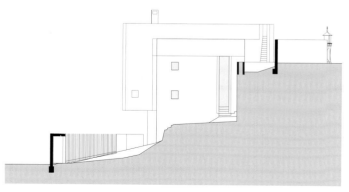

西立面 west elevation

项目名称：Garden House
地点：Haratzmüllerstraße 41, 4400 Steyr, Austria
建筑师：Hertl.Architekten ZT GMBH
项目团队：Gernot Hertl, Ursula Hertl
用地面积：790m²
总建筑面积：224m²
设计时间：2011
施工时间：2011
竣工时间：2014
摄影师：©Walter Ebenhofer(courtesy of the architect)

A-A' 剖面图 section A-A'

地下一层 first floor below ground

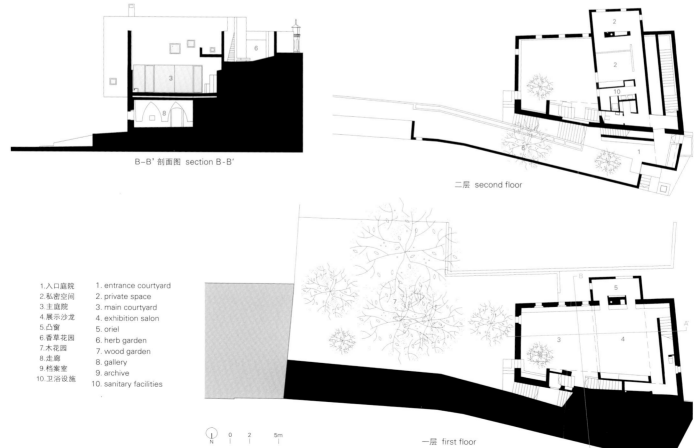

B-B' 剖面图 section B-B'

二层 second floor

1.入口庭院	1. entrance courtyard
2.私密空间	2. private space
3.主庭院	3. main courtyard
4.展示沙龙	4. exhibition salon
5.凸窗	5. oriel
6.香草花园	6. herb garden
7.木花园	7. wood garden
8.走廊	8. gallery
9.档案室	9. archive
10.卫浴设施	10. sanitary facilities

一层 first floor

time is open for guests who'd like to immerse into a place of being. We call it garden house because this describes the atmosphere of fusing intense green and space. The content is described by Refugium for retreating, Laboratorium for creative workshops, lectures and discussions, and Klausur for private viewings and small cultural events such as vernissages and concerts.

The old farmhouse is a ruin. The remaining outer walls create a large courtyard in which the new concrete volume is inserted. It's a house-in-house concept. Two patios are linked together beneath the small house into one space. Upstairs there are a guest room and the main entry. A staircase down to the level of the courtyard is hidden behind a porose new wall which doubles the outer wall. As a reference

light comes through small holes in the format of bricks. The new house cantilevers in the direction of the river Enns and creates an interior space outside the old walls. From here you can let your eyes wander across the water, where in the past the bank was used as a harbour for the rafts.

In terms of energy saving, resource optimisation and nature conservation issues, the sustainability of the building is a land-use-planning one, and it's based on using existing structures in the city. This should become the most important ecological answer for the whole Europe in future. To avoid constructing new buildings it saves much more of everything than any other scheme.

>>176
Hertl. Architekten ZT GmbH
Gernot Hertl was born in Steyr, Upper Austria in 1971 and graduated in 1997 from the Technical University in Graz. In 2003, he established Hertl. Architekten ZT GmbH. He received numerous architectural prizes including the International Architecture Award, Europe 40 Under 40 Award. Was nominated for the European Union Prize for Contemporary Architecture - Mies van der Rohe Award in 2015.

>>26
SchilderScholte Architects
Was founded by Gerrit Schilder[left] and Hill Scholte[right] in 2013. Gerrit Schilder is award winning double graduate and and has given lectures and workshops about his work and design theory at universities, institutes and symposiums worldwide. Through the years he has taught the way he approaches design to master's students at the Technical University of Delft, Avans University of Applied Sciences and the Piet Zwart Institute.
Hill Scholte was a jury member of the China Design Awards and was twice a honorable guest of the Guangzhou Design Week. Besides being the face of SchilderScholte Architects, she is also a course director for the bachelor Lifestyle Design of the Willem de Kooning Academy, University of Applied Sciences in Rotterdam.

>>38
Bosch Arquitectos
Ignacio Bosch and Borja Bosch are brothers born in Barcelona, Spain. They graduated in architecture from the University Polytechnic of Catalunya in Spain and worked for the studio Bosch Arquitectos, founded and directed by their father Javier Bosch. In 2012, they moved to Lima and set up Bosch Arquitectos together. They now focus on developing housing, facilities and urban design projects along Peru.

>>134
RCR Arquitectes
Rafael Aranda, Carme Pigem and Ramon Vilalta finished their studies in Architecture at the Superior Technical School of Architecture of Vallès in 1987. Since 1988, they have been working together as RCR Arquitectes in Olot, where they were born. They received National Award for culture in Architecture, 2005 in Catalonia, Chevalier de l'Ordre des arts et des lettres 2008 by the French government, Honorary Fellows by the American Institute of Architecture (AIA) 2010, Honorary Fellows by the Royal Institute of British Architects(RIBA) 2012 and Gold Medal by the French Royal Academy of Architecture 2015.

Aldo Vanini
Practices in the fields of architecture and planning. Had many of his works published in various qualified international magazines. Is a member of regional and local government boards, involved in architectural and planning researches. One of his most important research interests is the conversion of abandoned mining sites in Sardini.

>>90
Chiangmai Life Construction
Was founded in Northern Thailand, 2010 by Markus Roselieb. Tosapon Sittiwong, a young architect from northern Thailand joined in 2013. Since then, the company has specialized in modern earth and bamboo architecture and built several buildings. The company's philosophy is to adapt natural materials to the needs of the 21st century, so

>>38
AMA - Afonso Maccaglia Architecture
Was co-founded by Paulo Afonso and Marta Maccaglia in 2012. They have mainly developed educational projects in remote areas in the Peruvian central jungle. Paulo Afonso is graduated in architecture from the Architecture Department of Coimbra's University in Portugal. Since 2009, he has collaborated with several architecture studios including OAB-Carlos Ferrater Partnership in Barcelona, Spain alongside some independent work.
Marta Maccaglia has been working for international cooperation projects in Peru with NGOs since 2011. She founded the nonprofit Association "Semilla" in 2014, in which she is currently a director.

>>148
Irisarri-piñera Arquitectos
Is an architectural practice based in Vigo, Spain founded by Guadalupe Piñera Manso[left] and Jesús Irisarri Castro[right] in 1990. They are professors in the department of architectonics projects at architectural faculty in A Coruña, Spain. They also teach Sustainability and City at architectural faculty in Sevilla. They were invited to give courses and conferences for several universities and organizations both nationally and internationally. Their work have been published in some books and magazines, and they also wrote articles about many architectural projects.

Angelos Psilopoulos
Studied architecture at the School of Architecture, Aristotle University of Thessaloniki(AUTh), then moved on to his Post-Graduate studies at the National Technical University in Athens(NTUA). Is currently pursuing his Ph.D. at the NTUA on the subject of Theory of Architecture, studying gesture as a mechanism of meaning in architecture. Has been working as a freelance architect since 1998, undertaking a variety of projects both on his own and in collaboration with various firms and architectural practices in Greece. Since 2003, he has been teaching Interior Architecture and Design in the Department of Interior Design, Decoration, and Industrial Design at the Technological Educational Institute of Athens(TEI).

>>102
Dekleva Gregorič Arhitekti
Was founded in 2003 by Aljosa Dekleva[right] and Tina Gregorič[left] in Ljubljana, Slovenia. Both graduated from University of Ljubljana, Slovenia and continued their study at the Architectural Association School of Architecture where they received Master degree in Architecture with distinction in 2002. They have been visiting lecturers and critics at the AA School, Technical University of Graz, Austria and many others. Understand design as research on several modes (social, material, and historical) and respond to specific constrains and conditions. Rather than a conceptual approach to intense structuring of space, they challenge the use of materials and exposure their natures. They are aiming to stimulate new social interactions among users, participation of the users, design process and customization to users' needs.

>>56
Toshiko Mori Architect
Toshiko Mori was born in 1951 and graduated from Cooper Union School of Architecture in 1976. She worked for Edward Larrabee Barnes and after leaving the firm, in 1981 she founded Toshiko Mori Architect in New York. Her intelligent approach to ecologically sensitive siting strategies, historical context, and innovative use of materials reflects a creative integration of design and technology. Her designs demonstrate a thoughtful sensitivity to detail and involve extensive research into the site conditions and surrounding context.

>>46
Kéré Architecture
Diébédo Francis Kéré was born in Burkina Faso and graduated from the Technical University of Berlin, Germany in 2004. Founded Kéré Architecture in 2005 and Schulbausteine für Gando (Bricks for Gando), a charitable foundation focused to realize sustainable architecture for the community. He develops strategies for innovative construction by combining traditional techniques and materials with modern engineering methods. Today, he continues to support the educational, cultural, and sustainable needs of communities in Burkina Faso.

>>70
dEEP Architects
Li Daode was born in 1981 and graduated from the Central Academy of Fine Arts in Beijing and the Architectural Association School of Architecture. Worked for Norman Foster in his London office until he establishes his own office in Beijing in 2009. dEEP Architects is committed to use advanced design concept and high-tech design method to integrate with the reality of the local construction and the ongoing urbanism in China. They also work on different scales of design projects which includes planning, architecture, landscape, interior and product design.

>>114
Budi Pradono Architects
Budi Pradono was born in Salatiga, Indonesia and graduated with distinction from Duta Wacana Christian University, Architecture Department in Yogyakarta. In 1999, he established Budi Pradono Architects. After that, he received M.Arch at the Berlage Institute Post Graduate Laboratory of Architecture in Rotterdam. He has received several awards including Prize IAI Jakarta Awards. From June 2015 to April 2016, his exhibition "The Bathroom of the 1970s" is being held in Schiltach, Germany.

>>80
Kikuma Watanabe
Was born in 1971, Japan. Graduated from Department of Architecture, Faculty of Engineering at Kyoto University in 1994 and completed doctoral program at the same department in 2001. In 2007, he founded D Environmental Design System Laboratory. He is currently teaching as an associate professor at Kochi University of Technology and he is also engaged in research and fieldwork in various regions.

>>162
Wespi de Meuron Romeo Architects
Was founded by Markus Wespi[left], Jérôme de Meuron[middle] and Luca Romeo[right] in 2012. Markus Wespi was born in 1957 in St. Gallen, Switzerland. He had learned architecture for himself. Has been working with Jérôme de Meuron since 1998. Jérôme de Meuron was born in 1971 in Münsingen, Germany and studied at the Technical School of Burgdorf from 1993 to 1996. He experienced architectural practical while staying in Ghana, Africa from 1996 to 1997. Luca Romeo was born in 1984 in Locarno, Switzerland and studied at the Technical School of Lugano-Trevano from 2003 to 2006. He has been working with Markus Wespi and Jérôme de Meuron since 2011.

C3, Issue 2015.12
All Rights Reserved. Authorized translation from the Korean-English language edition published by C3 Publishing Co., Seoul.

© 2016 大连理工大学出版社
著作权合同登记06-2016年第99号

版权所有·侵权必究

图书在版编目(CIP)数据

地方性与全球多样性：汉英对照 / 韩国C3出版公社编；张琳娜等译. — 大连：大连理工大学出版社，2016.7

(C3建筑立场系列丛书 / C3出版公社)

书名原文：C3: Regionalism and Global Diversity

ISBN 978-7-5685-0454-6

Ⅰ. ①地… Ⅱ. ①韩… ②张… Ⅲ. ①建筑设计—汉、英 Ⅳ. ①TU2

中国版本图书馆CIP数据核字(2016)第168130号

出版发行：大连理工大学出版社
　　　　　（地址：大连市软件园路80号　邮编：116023）
印　　刷：上海锦良印刷厂
幅面尺寸：225mm×300mm
印　　张：11.75
出版时间：2016年7月第1版
印刷时间：2016年7月第1次印刷
出 版 人：金英伟
统　　筹：房　磊
责任编辑：杨　丹
封面设计：王志峰
责任校对：周小红
书　　号：978-7-5685-0454-6
定　　价：228.00元

发　行：0411-84708842
传　真：0411-84701466
E-mail：12282980@qq.com
URL：http://www.dutp.cn